Rereading Darwin's *Origin of Species*

Explorations in Science and Literature

Series Editors:
John Holmes, Anton Kirchhofer and Janine Rogers

Explorations in Science and Literature consider the significance of literature from within a scientific worldview and bring the insights of literary study to bear on current science. Ranging across scientific disciplines, literary concepts, and different times and cultures, volumes in this series will show how literature and science, including medicine and technology, are intricately connected, and how they are indispensable to one another in building up our understanding of ourselves and of the world around us.

Published titles
Biofictions, Josie Gill
Imagining Solar Energy, Gregory Lynall
The Diseased Brain and the Failing Mind, Martina Zimmermann
Narrative in the Age of the Genome, Lara Choksey
Writing Remains, Edited by Josie Gill, Catriona McKenzie, Emma Lightfoot

Forthcoming titles
Entwined Being, Edward King
The Social Dinosaur, Will Tattersdill

Rereading Darwin's *Origin of Species*

The Hesitations of an Evolutionist

Richard G. Delisle and James Tierney

BLOOMSBURY ACADEMIC
LONDON • NEW YORK • OXFORD • NEW DELHI • SYDNEY

BLOOMSBURY ACADEMIC
Bloomsbury Publishing Plc
50 Bedford Square, London, WC1B 3DP, UK
1385 Broadway, New York, NY 10018, USA
29 Earlsfort Terrace, Dublin 2, Ireland

BLOOMSBURY, BLOOMSBURY ACADEMIC and the Diana logo are
trademarks of Bloomsbury Publishing Plc

First published in Great Britain 2022
This paperback edition published 2023

Copyright © Richard G. Delisle and James Tierney, 2022

Richard G. Delisle and James Tierney have asserted their right under the Copyright,
Designs and Patents Act, 1988, to be identified as the Authors of this work.

For legal purposes the Acknowledgments on p. xii constitute
an extension of this copyright page.

Cover design: Rebecca Heselton
Cover image © Portrait of Charles Darwin by Julia Margaret Cameron
(Photo by © Historical Picture Archive/CORBIS/Corbis via Getty Images)

All rights reserved. No part of this publication may be reproduced or
transmitted in any form or by any means, electronic or mechanical, including
photocopying, recording, or any information storage or retrieval system,
without prior permission in writing from the publishers.

Bloomsbury Publishing Plc does not have any control over, or responsibility for,
any third-party websites referred to or in this book. All internet addresses given
in this book were correct at the time of going to press. The author and publisher
regret any inconvenience caused if addresses have changed or sites have
ceased to exist, but can accept no responsibility for any such changes.

A catalogue record for this book is available from the British Library.

A catalog record for this book is available from the Library of Congress.

ISBN:	HB:	978-1-3502-5957-7
	PB:	978-1-3502-5976-8
	ePDF:	978-1-3502-5958-4
	eBook:	978-1-3502-5959-1

Series: Explorations in Science and Literature

Typeset by Integra Software Services Pvt. Ltd.

To find out more about our authors and books visit www.bloomsbury.com
and sign up for our newsletters.

*I dedicate this book to the memory of
Jacob Bronowski (1908–1974), whose BBC television series*
The Ascent of Man *ignited the imagination of a very young scholar.*

*His praise for England's tolerance regarding scientific ideas moved me,
a reputation manifestly not overrated as seen in the
publication of this book in that country.*

Richard G. Delisle

I dedicate this book to my mother and father, Ann and James Tierney.

James M. Tierney

Contents

About the Authors	viii
Preface	ix
Acknowledgments	xii
Introduction: The Two Sides of Darwin	1

Part One The Charles Darwin We Think We All Know

1	A Primer of Evolution's Complexities	13
2	What Time Selected from Darwin: The Standard View	45

Part Two Charles Darwin and the Static Worldview

3	The Tree That Hides the Forest: Charles Darwin's "Tree of Life"	63
4	Divergence: A Geometry That Shatters Creative Time and Novelty	89
5	A Cyclical World in Equilibrium	99
6	Natural Selection: The Core of Darwin's Theory?	121

Part Three Charles Darwin Viewed in Piecemeal Fashion

7	When So-Called New Ideas Hide Old Ones	137

Conclusion: Back to the Future	145
Index	158

About the Authors

Richard G. Delisle holds a PhD in paleoanthropology and a PhD in philosophy. He is Associate Professor at the University of Lethbridge (Canada) where he teaches evolution and history/philosophy of science at the School of Liberal Education, being also affiliated with the Department of Philosophy. He is the founder and editor of the book series "Evolutionary Biology: New Perspectives on its Development". He is the author of *Debating Humankind's Place in Nature, 1860–2000: The Nature of Paleoanthropology* (2007) and *Charles Darwin's Incomplete Revolution: The Origin of Species and the Static Worldview* (2019), among other publications.

James Tierney studied philosophy and French at the University of Michigan and philosophy at the University of Chicago. He is currently Senior Lector and Director of Yale English Language Programs. As part of the founding board of the Consortium on Graduate Communications, he organized its first Summer Institute at Yale University (in 2016) and is active in research and advocacy in the field of advanced language learning at the graduate level. He has also worked as a freelance editor and translator since 2005.

Preface

Written for a general audience, this book attempts to think through the *Origin of Species* with Darwin and, in so doing, invite the reader to discover neglected ambiguities and contradictions in Darwin's magnum opus. Whether or not one accepts the book's admittedly controversial conclusions, we hope the journey will be rewarding in itself, since the *Origin* is undoubtedly a work of profound importance that repays consideration from many perspectives. It is through precisely such an examination and subsequent (at times agonizing!) reappraisals that this book came to be written.

For more than ten years, RGD, the book's lead author, has been teaching a yearly university course entirely devoted to the *Origin*. To this day, this course is taught in a manner wholly faithful to the received view, one that places Darwin front and center and that casts him as the first truly modern evolutionist. Moving in parallel with that teaching, RGD's research began to arouse suspicions regarding doctrinaire interpretations of Darwin eight years ago. The immediate context for these worries was an attempt to get to grips with a fairly innocuous aspect of Darwin's thinking: the well-known fact that Darwin did not collect much evidence from the fossil record to support his theory. Indeed, Darwin had at best a complicated relationship with the field of paleontology, preferring instead to build his case by identifying traces of evolution within currently existing forms. This earlier research sought to gauge how this particular methodological bias may have affected Darwin's view of evolution, guided by the working hypothesis that this might have had a distorting effect on his conception of the evolutionary past: nothing more dramatic than this. What began as an apparent side issue, however, became ever more gripping and has, it must be said, loomed ever more threatening to the Darwin Legend, with each passing year of research. That said, the conclusions to be presented here were largely unexpected at the outset and came about largely accidently, following a fairly conventional investigative approach through to its conclusions.

For JT, the book's second author, the work has also meant a reassessment of Darwin, but approached from the perspective of one interested in the history and philosophy of science. His involvement in the project evolved out of a long partnership with RGD as an editor, interlocutor, and friend, absorbing and discussing the many issues considered here. A philosopher by training, he is drawn to questions of humanity's place in nature, including its relation to its past and future, matters which are in his view insufficiently contemplated in our scientific and technological age. While his admiration for Darwin and his remarkable achievements has never wavered, he has long had qualms about recent hagiographic treatments of the man and his work, and his elevation to iconic status. The project was thus a welcome opportunity to engage with Darwin's thought from a more critical angle.

Through his work the *Origin of Species*, Charles Darwin (1809–82) is revered as a revolutionary thinker who brought us nearly single-handedly to confront the truth of evolution. He is remembered for having launched a new epoch, for providing us our modern and seemingly unshakeable understanding of humanity's past, as well as its relation to nature. For the wider public, Darwin is the founding figure of evolution; for many evolutionists, he is the first to have proposed a truly modern theory of evolution by providing it with a firm foundation upon which they later built the tremendous edifice of evolutionary biology. Nonetheless, as is so often the case with historical figures, Darwin's greatness seems to have obscured the man and his work to some extent: both friends and enemies have mischaracterized his work.

This book, written by two scholars having no quarrel with Darwin, strives to present a more comprehensive picture of his ideas. It begins by asking a simple and perhaps naive-sounding question: how could a mid-nineteenth-century scholar like Darwin have managed to present, nearly single-handedly, a modern evolutionary theory, while many of his contemporaries were still struggling to transition out of a conception of nature and time as essentially fixed and unchanging? Pressing this question yielded a surprising answer: Darwin, too, struggled to make this transition and was in fact not entirely successful in extricating himself from this older worldview and other widely held ideas that formed his intellectual milieu. For those willing to look, one finds at the core of the *Origin of Species* a mixed bag of intellectual commitments deeply indebted to seventeenth- and eighteenth-century thought, ideas ill-fitted to support anything like our modern notion of evolutionism.

Does this mean that Darwin's *Origin* is "much ado about nothing"? Of course not; it remains a remarkable work, but it is also a work of its time. Returning Darwin to the human dimension allows us to set aside the nearly mythical status at times awarded to him, consciously or unconsciously, by overzealous enthusiasts who would lionize him and religious opponents who would demonize him. This, if anything, should only *increase* our esteem for Darwin and his work.

Secure in our own evolutionary worldview, it is easy to pass over the enormity of the intellectual challenge accompanying what is perhaps the most profound revolution in thought during the last centuries, the shift from a static to an evolutionary worldview. Unable to break free from this older outlook entirely, Darwin offered a theory that, ultimately, constitutes a fascinating compromise between the old and the new. Acknowledging this as a *compromise* does nothing to diminish his achievement: it simply means that one must be careful in how one approaches the notion of *revolution* in the history of science. Rediscovering this other Darwin—and this other side of the *Origin of Species*—also helps us grasp the immensity of the task that lay before Darwin and other nineteenth-century scholars, and as well as their ultimate achievements.

While the approach to Darwin's *Origin of Species* to be presented here departs sharply from other recent accounts, something we make no bones about, we also hope to provide readers with a strong sense of the evolutionary landscape. Before turning to the arguments of this book, then, we shall begin with something less contentious, orienting the reader by presenting key issues at play in evolutionary studies today. For readers interested in delving deeper into particular issues encountered along the way, we have included a detailed annotated bibliography at the conclusion of several key chapters, which present, among other things, recent developments in both Darwin studies and evolutionary biology.

Acknowledgments

We would like to express our gratitude to our dedicated editor, John Holmes, who elegantly accompanied us and this book to its final destination, and for his many helpful suggestions. We also extend our recognition to two anonymous reviewers whose much-appreciated recommendations definitely improved this book.

Introduction: The Two Sides of Darwin

In 2014, the prestigious scientific journal *Nature* opened up its pages to the following debate: Does evolutionary theory need a rethink? While one group of scientists answered, "Yes, urgently," another group hastily replied, "No, all is well." It is the response of this latter group that is of interest to us, given the kind of justification they offered: "The evolutionary phenomena championed by [scientists requesting an urgent rethink] are already well integrated into evolutionary biology, where they have long provided useful insights. Indeed, all of these concepts date back to Darwin himself."[1] This is far from an isolated case; it is indeed common among evolutionists today to take for granted that the work produced by Charles Darwin in the mid-nineteenth century already constituted a firm and solid basis on which to construct the edifice of evolutionary biology. In his now-classic *Sociobiology* (1975), Edward O. Wilson writes: "Sociobiology will perhaps be regarded by history as the last of the disciplines to have remained in the 'unknown land' beyond the route charted by Darwin's *Origin of Species*."[2] In an emotional passage in his magnum opus published at the very end of his life, *The Structure of Evolutionary Theory* (2002), Stephen Jay Gould writes:

> When I ask myself how all these disparate thoughts and items fell together into the one long argument of this book, I can only cite ... my love of Darwin and the power of his genius. Only he could have presented such a fecund framework of a fully consistent theory, so radical in form, so complete in logic, and so expansive in implication. No other early evolutionary thinker ever developed such a rich and comprehensive starting point [allowing us to leave] ... Darwin's foundation intact while constructing a larger edifice of interestingly different form thereupon.[3]

Glowing praise indeed! But it is, we shall argue, a hyperbole typical of writings on Darwin. After all, it is one thing to acknowledge that, particularly in comparison with other thinkers of his day, Darwin left us with "a rich and comprehensive" place from which to carry on the work of evolutionary study; it is quite another to suggest that his heirs carried on his thought "intact" as Gould suggests and many self-proclaimed Darwinians would have us believe. Here we see the Darwin Myth in a nutshell: the isolated, protean genius ("only he could have presented") from whom sprung fully formed a solid scientific theory ("a fully consistent theory, so radical in form, so complete in logic"), which, for all practical purposes, is the foundation of modern evolutionary thought ("a comprehensive starting point").

Is it really reasonable to think that Darwin single-handedly initiated a wholesale and definitive shift in biological thought in the *Origin*? Another eminent evolutionist, Ernst Mayr, put it this way in 2001: "The actual shift from the belief in a static worldview to evolutionism was caused by the dramatic event of the publication of Charles Darwin's *On the Origin of Species* on the 24th of November in 1859."[4] In *The Greatest Show on Earth: The Evidence for Evolution* (2009), Richard Dawkins comes quite close to stating the same idea: "By the time Darwin came to publish *On the Origin of Species* in 1859, he had amassed enough evidence to propel evolution itself ... a long way towards the status of fact."[5]

One might be tempted to view such statements by eminent scholars as appropriate displays of modesty, giving Darwin his due by acknowledging that in their own work they have, like Newton, "stood on the shoulders of giants" (or in this case *a giant,* in the singular). No doubt this was, in part at least, their intent. Nonetheless, the net effect of tying one's work to the Master cannot be denied, perhaps sometimes even concealing a substantial redefinition of an intellectual project, in this case Darwinism itself. As Eva Jablonka and Marion J. Lamb have very recently observed in characterizing these redefinitions promulgated by more recent evolutionists: "This was certainly not Darwin's Darwinism—It was a version of Neo-Darwinism, but labelling this view as 'Darwinism' undoubtedly endowed it with more authority."[6]

This book will expand upon the contrast that should be established between Darwin himself and self-proclaimed "Darwinians" of the twentieth century and after. In science as in other domains, it is often the case that new perspectives emerge only at the cost of asking daring questions, questions that enable us to shake off the weight of the tradition. Let us begin by asking an admittedly very strange question: "Was Charles Darwin really an evolutionist?" Of course, putting the question this way immediately gives rise to a second question: Is the person asking it a religious zealot, crazy, desperate for media attention or internet notoriety? We assure you these reasons do not apply in our case, and admit that this question sounded no less crazy to us only a short while back. After all, as most of us were taught from an early age in science class, Darwin is the "founding figure" of evolution, his name nearly synonymous with it, and the *Origin of Species*, first published in 1859, is its founding document. If there were ever an unshakeable edifice in the history of science, it would be Darwin and the achievement the *Origin* certainly is.

Much has been built on this foundation: it is no exaggeration to speak of a "Darwin Industry" as we do today, in both the scholarly and popular literature. Darwin is idolized by many as quite literally an *iconoclast* ("a destroyer of religious pictures"), a hero of reason and the scientific spirit. Quite understandably, he is despised by others, for the very same reasons. For admirers and detractors both, he is assumed in the popular literature to be the progenitor of an idea called "evolution." Lurking behind the "Darwin Legend" lies the hazy image of a bearded genius sailing off to exotic isles to look at finches and turtles, in the process doing nothing less than changing our understanding of the world (and our place in it) through his diligent investigations.

While much of science lies in the art of discovery and explanation, the scientific mind is also a critical and skeptical one, trained and ready to question assumptions and to shake foundations once thought unassailable: Darwin himself is a prime example of this questioning spirit and a worthy role model in this regard, having dared to challenge some assumptions that threatened not only the scientific thinking of his day, but also a deeply entrenched cultural and religious order. It is thus very much in the scientific spirit to ask whether much of what has been written about him, and in a sense, *in his name,* is in fact true or rightly credited to him. And this is far from crazy, since once one moves beyond a superficial level and looks carefully at the writings themselves, one sees that, despite herculean efforts, Darwin was only partially able to exorcize the ghosts of the past, often retaining antiquated notions that permeated the thought of his day. While perhaps unsurprising in itself, it is worth noting that Darwin unknowingly imported many *anti*-evolutionary ideas into his work, in particular what we will here term a "static worldview"—an understanding of the world as fully formed and complete right from the very start, a world deprived of a past different from what can be seen today—a notion inimical to evolutionary thought as we understand it today.

As a culture, we are swiftly moving past appeals to "founding figures" when attempting to talk about our history; as historians, we have long ago set aside accounting for movements of thought solely in terms of the contributions of "great actors." Indeed, someone's having acquired this status is typically reason enough to start asking tougher questions. While this book is not intended as a "takedown" of the scholar of the Galapagos, it is hoped it will open some cracks in an image of Darwin that is in many ways built on dubious foundations. Only when one strips away the rhetoric, and looks closely at the arguments, claims, and assumptions of Darwin's *Origin of Species*, and importantly, as one isolates the presuppositions those of us schooled in more modern scientific traditions bring to the book, a picture emerges not of an isolated revolutionary, but of someone very much of his time, who shared much of the thinking of other less-acclaimed figures such as his contemporaries and countrymen Charles Lyell, Richard Owen, and Thomas Henry Huxley, all of whom were slowly inching their way toward evolutionism in the 1850s and 1860s.

The *Origin of Species* is a deceptively difficult book, not so much because of its writing style and apparent clarity of exposition, but because of Darwin's subtly convoluted way of laying out his arguments and facts, filling gaps in knowledge by making extensive use of imaginary scenarios and thus concealing contradictions with argumentative strategies that tend to dissimulate hidden, yet, significant commitments. Speaking of the use of imaginary scenarios in the *Origin*, Devin Griffiths writes: "[Darwin] commonly asks the reader to imagine something with him ... that is effectively invisible ... He uses imaginative history to explore how stories about what might have happened explain what we find in nature."[7] As for argumentative inconsistencies, these did not go unnoticed by his contemporaries. In 1958, nearly one hundred years after the original publication of the *Origin*, Alvar Ellegard surveyed the reception of Darwin's magnum opus following its immediate publication. He thus summarized his survey: "A host of press voices accused Darwin of mingling together 'facts and assumptions, probabilities and speculations .... in most illogical confusion' ... and of using his imagination to supply the lack of factual observation."[8] In Chapter 2 we will present similar comments from scholars confronting Darwin's work and legacy. The difficulties raised by Ellegard and others suggest a closer look at Darwin's overall argumentative strategy is not out of place.

"If one is ready to read Darwin as a writer rather than a twenty-first-century evolutionary biologist,"[9] George Levine observes, one is likely to discover that his writing is permeated with sensibilities deeply rooted in his own time, the Victorian era. In his words, "unequivocal cooptations"[10] aptly describe the practice of some recent evolutionists when appropriating Darwin's work, distorting the true nature of his contribution. Studies of science and the ways scientific discourses are elaborated have progressed far enough to have seen the rise of a more critical approach, and thus of critics devoted to exploring the "sociology of scientific knowledge" wherein it is recognized that the demarcation line between science per se and the context in which it is conducted is quite porous, if traceable at all. As John Holmes notes, "[f]or such critics science was very much part of culture, and culture was best understood in its historical context."[11] In order to examine some of what can be found in the textual fabric of the *Origin of Species*, it is useful to turn, at this early stage of our investigation, to Gillian Beer who explains in *Darwin's Plots*:

> Darwinian theory will not resolve to a single significance nor yield a single pattern. It is essentially multivalent. It renounces a Descartian clarity, or univocality. Darwin's methods of argument and the generative metaphors of *The Origin* lead … into profusion and extension. The unused, or uncontrolled, elements in metaphors … take on a life of their own. They surpass their status in the text and generate further ideas and ideologies. They include 'more than the maker of them at the time knew'.[12]

While Beer highlights both the ambivalence and the fecundity of the work, this assessment also raises the pressing question of what is actually there and what has been merely superimposed upon it. The ambiguity she cites, quite correctly in our view, constitutes a feature many later scholars have found irresistible to remedy on Darwin's behalf in their own interpretations.

Our book is dedicated to digging beneath the surface of what the reader first encounters in the *Origin*, trying to work our way through the layers of Darwin's thinking, perhaps even into unconscious territory. Whereas we acknowledge that recent studies produced within the context of "sociology of scientific knowledge" have opened the door to such a critical approach, we also think it is possible to go one step further by showing that Darwin was even more deeply indebted to his own historical period than has thus far been acknowledged. In particular, it is still widely held in spite of such critical studies that Darwin initiated a profound intellectual break, a true revolution in the history of ideas. We hold, on the contrary, that his ideas concorded much more smoothly with the intellectual currents of his time, following a course that guided scholars through a gradual transition between the static and truly evolutionary worldviews, leaving it to others to complete this revolution.

Reassessing Charles Darwin's significance for the modern scientific tradition—one rivaled only, perhaps, by that of Albert Einstein—is a large part of our motivation in sharing this alternative outlook. So-called founding figures cannot be the private property of specialized scholars or scientists in open democratic societies, for they are also *public* figures belonging to all. The oversimplified picture of Darwin and his work typically presented to the public is partial and incomplete. Once the complexities underlying the long and convoluted transition from a static worldview to modern evolutionary thinking are taken into account, it becomes much harder to accept that a single iconic figure might serve as the sole and definitive incarnation of modern evolutionism.

This is the sobering lesson that we have drawn from the experience of unlearning what we thought we knew about the *Origin of Species*. It must be remembered that Darwin was no evolutionist in the late 1820s and early 1830s, although he was beginning to be intrigued by the idea at that time. Is it reasonable simply to assume, then, that by 1859 he had managed to modify his thought to such a point that he might rightly be considered the first truly modern evolutionist? It seems more likely that, along with other scholars of his time, Darwin had embarked on an intellectual transition between the static worldview and the fully modern evolutionary one.

Our surprising claim raises a question: Why has it taken so long for us to realize that Darwin's commitment to evolutionism was incomplete? It seems that a combination of four factors is involved:

1. In the mid-nineteenth century, scholars were less concerned with distinguishing between the various brands of evolutionism than on drawing a clear demarcation line between anti-evolutionists and evolutionists. The precise nature of Darwin's own transition to evolutionism was, therefore, concealed from his contemporaries.
2. The *Origin of Species* is a truly complex book fusing many levels of analysis and full of interpretative pitfalls. Putting order in it constitutes a challenge, for Darwin's contemporaries as for us today.
3. Over time, scholars have tended to project their own modern understandings of evolution onto Darwin's book, thus making it difficult to clearly see what's in it.
4. The transition to evolutionism was not a spontaneous change of course in the history of ideas. It has been a slow and complex process that is still under way. We now have a much more profound understanding of the implication accompanying evolutionism than the evolutionists of the mid-twentieth century possessed only a short time ago. It is this gradual process of deepening of our understanding that, perhaps, will today allow us to gauge the extent to which Charles Darwin was no modern evolutionist.

Two caveats are in order before tackling our main topic. First, what is said of Charles Darwin here concerns only what can be found in the *Origin*. We will not take up Darwin's other publications, such as *The Variation of Animals and Plants under Domestication* (1868) and *The Descent of Man* (1871). A complete reevaluation of Darwin's work is a more ambitious, collaborative, and longer-term undertaking. Nonetheless, it should be kept in mind that, with no fewer than six editions between 1859 and 1872, the *Origin of Species* likely represents the single most important work of Darwin's mature years (he died in 1882) and is rightly considered a definitive statement of his thought. Second, while authors should not be held responsible for the use or misuse of their work by others, we are also well aware that any work perceived as seeking to take Darwin down a few pegs is likely to appeal to opponents of evolution. This would be a mistake in our case, however, since the re-examination of Darwin we undertake here in no way threatens the truth of evolution, at this point a well-established and incontrovertible fact.

Annotated Bibliography

For a comprehensive view of the 2014 debate published in the journal *Nature*, compare: Laland, K., T. Uller, M. Feldman, K. Sterelny, G. Müller, A. Moczek, E. Jablonska, and J. Odling-Smee (2014), "Does Evolutionary Theory Need a Rethink? Yes, Urgently," *Nature*, 514: 161–4 and Wray, G., H. Hoekstra, D. Futuyma, R. Lenski, T. Mackay, D. Schluter, and J. Strassmann (2014), "Does Evolutionary Theory Need a Rethink? No, All Is Well," *Nature*, 514: 161–4.

It is worth noting that not all evolutionists today aim to build evolutionary biology upon the views of Charles Darwin. For anti-Darwinian theories, consult, for instance: Reid, R. G. B. (1985), *Evolutionary Theory: The Unfinished Synthesis*, Croom Helm: Beckenham; Kauffman, S. (1993), *The Origins of Order: Self-Organization and Selection in Evolution*, Oxford: Oxford University Press; Goodwin, B. (1994), *How the Leopard Changed Its Spots: The Evolution of Complexity*, Princeton: Princeton University Press; Ulanowicz, R. (2009), *A Third Window: Natural Life beyond Newton and Darwin*, West Conshohocken, PA: Templeton Foundation Press. This being said, it is also the case that the overwhelming number of evolutionists who see themselves as building upon Darwin's work are doing so under the assumption that his biology was already remarkably modern. We argue, on the contrary, that many explanatory components found in the *Origin of Species* were conceptualized by him in a rather unmodern way.

Readers interested in learning more about the life of Charles Darwin are invited to consult, among a long list of available works, the following: Desmond, A. and J. Moore (1991), *Darwin: The Life of a Tormented Evolutionist*, New

York: W.W. Norton; Browne, J. (1995), *Charles Darwin: Voyaging*, New Jersey: Princeton University Press; Browne, J. (2002), *Charles Darwin: The Power of Place*, New Jersey: Princeton University Press; Desmond, A. and J. Moore (2009), *Darwin's Sacred Cause*, Boston: Houghton Mifflin Harcourt. Many other references will be provided over the course of this book, with an emphasis on those taking up Darwin's scientific works.

The works of Charles Darwin have been studied for quite some time. However, beginning in the 1950s this research activity greatly intensified, first with a closer study of all his published works, followed by a careful examination of his notebooks, correspondence, and even marginalia. In terms of historiographical approach, we adopt a "back to basics" approach: rereading a single piece of work, the *Origin of Species*, with even greater care and with the benefit of the fresh perspectives afforded to us by the deeper understanding of evolutionism we have today.

We are hardly the first to try to place Charles Darwin within his own historical context and to consider his intellectual roots in the eighteenth and preceding centuries. Significant efforts have long been under way trying to do just that, as can be seen in this small sample of works in this vein: Greene, J. C. (1981), *Science, Ideology, and Word View*, Berkeley: University of California Press; Richards, R. J. (2004), "Michael Ruse's Design for Living," *Journal of the History of Biology*, 37: 25–38; Ruse, M. (2004), "The Romantic Conception of Robert J. Richards," *Journal of the History of Biology*, 37: 3–23; Hodge, M. J. S. and G. Radick (2009), "The Place of Darwin's Theories in the Intellectual Long Run," in M. J. S. Hodge and G. Radick (eds), *The Cambridge Companion to Darwin*, 2nd edn, 246–73, Cambridge: Cambridge University Press.

As will be seen in the following chapters, this book sides in a significant way with the interpretative tradition that emphasizes the impact of "natural theology" on Darwin's way of thinking, an idea long recognized by historians and philosophers in the secondary literature. However, our own understanding moves away from the classic interpretation by claiming that the impact of natural theology on Charles Darwin is accompanied by somewhat concealed, yet profound, commitments toward notions more properly articulated in the wider intellectual context provided by the *static worldview*. In this sense, "natural theology" constitutes, for Darwin, something like the tip of the iceberg. These commitments will be gradually unpacked as this book unfolds. For greater detail on our position as it relates to the traditional historiography, see Delisle, R. G. (2019), *Charles Darwin's Incomplete Revolution: The Origin of Species and the Static Worldview*, Switzerland: Springer Nature, 1–30.

It is hard to exaggerate the complexity and difficulty of the transition, beginning in the eighteenth century, from a static to an evolutionary worldview. This gave rise to all sorts of compromises and intermediary positions lying between these two perspectives. An excellent and very readable account of this transition during the eighteenth and nineteenth centuries is provided in Bowler, P. J. (2003), *Evolution: The History of an Idea*, 3rd edn, Berkeley: University of California Press. What is missing from this account, in our view, is the extent to which Darwin was also committed to a compromise position between the two worldviews. Also missing is the notion that "evolution" itself had not yet reached its full maturity during the nineteenth century, a maturity process still under way today. Indeed, the idea of "evolution" encapsulates more than the notion of "evolutionary change." Whereas "change" is part of the package, it also extends to include, for instance, a methodological and ontological reflection on the relationship between the past and the present. This reflection will be fleshed out in our Chapter 2.

For an analysis of the lack of a clear demarcation line between science itself and the society that makes it possible, consult Shapin, S. (1992), "Discipline and Bounding: The History and Sociology of Science as Seen through the Externalism-Internalism Debate," *History of Science*, 30: 333–69. For excellent overviews of what we have here referred to as the "sociology of scientific knowledge," including key references and the limitations of the field, consult Shapin, S. (1995), "Here and Everywhere: Sociology of Scientific Knowledge," *Annual Review of Sociology*, 21: 289–321; Zammito, J. H. (2004), *A Nice Derangement of Epistemes: Post-Positivism in the Study of Science from Quine to Latour*, Chicago: University of Chicago Press.

Scholars working in many different fields have attempted to understand Charles Darwin's work; these include evolutionary biologists, traditional historians and philosophers of science, and scholars inspired by the sociology of scientific knowledge. As might be expected, approaching Darwin from different perspectives may lead to conflicting assessments, raising the issue of a "harmonious body of knowledge" or "consilience." For discussion of this issue, consult Holmes, J. (2009), *Darwin's Bards*, Edinburgh: Edinburgh University Press, 1–22. Our book will expand upon what George Levine calls "unequivocal cooptations" (see main text of this chapter). We would like to point out that scholars of various stripes may have uncritically assimilated what leading evolutionary biologists have said about Darwin's work, leading to a premature and somewhat superficial "consilience." For such a critical assessment, consult Delisle, R. G. (2021), "Introduction: In Search of a New Paradigm for the Development of Evolutionary Biology," in R. G. Delisle (ed), *Natural Selection: Revisiting Its Explanatory Role in Evolutionary Biology*, 1–8, Switzerland: Springer Nature. While scholars inspired by the sociology of scientific knowledge have already detected a number of inconsistencies that throw traditional interpretations of Darwin's work into question, it seems many have stopped short of tracing out their implications for Darwin studies more generally (perhaps under the weight of the interpretive tradition). We will return to this issue throughout the remainder of the book, including in our annotated bibliographies.

Notes

1. Wray, G., H. Hoekstra, D. Futuyma, R. Lenski, T. Mackay, D. Schluter and J. Strassmann (2014), "Does Evolutionary Theory Need a Rethink? No, All Is Well," *Nature*, 514: 163.
2. Wilson, E. O. (1975), *Sociobiology: The New Synthesis*, Cambridge MA: Belknap Press, 63.
3. Gould, S. J. (2002), *The Structure of Evolutionary Theory*, Cambridge MA: Belknap Press, 47.
4. Mayr, E. (2001), *What Evolution Is*, New York: Basic Books, 9.
5. Dawkins, R. (2009), *The Greatest Show on Earth: The Evidence for Evolution*, New York: Free Press, 18.
6. Jablonka, E. and M. J. Lamb (2020), *Inheritance Systems and the Extended Evolutionary Synthesis*, Cambridge: Cambridge University Press, 8.
7. Griffiths, D. (2016), *The Age of Analogy: Science and Literature between the Darwins*, Baltimore: Johns Hopkins University Press, 11–12.
8. Ellegard, A. ([1958] 1990), *Darwin and the General Reader: The Reception of Darwin's Theory of Evolution in the British Periodical Press, 1859–1872*, Chicago: University of Chicago Press, 189.
9. Levine, G. (2011), *Darwin the Writer*, Oxford: Oxford University Press, 11.
10. Levine, G. (2009), "Foreword," in G. Beer, *Darwin' Plots*, 3rd edn, Cambridge: Cambridge University Press, xi.
11. Holmes, J. (2009), *Darwin's Bards*, Edinburgh: Edinburgh University Press, 1.
12. Beer, G. (2009), *Darwin' Plots*, 3rd edn, Cambridge: Cambridge University Press, 6–7.

Part One

The Charles Darwin We Think We All Know

1

A Primer of Evolution's Complexities

Before taking up Charles Darwin's intellectual achievements, it will be useful for us to review a few of the key elements of evolutionary theory in its contemporary form. This may serve to refresh the memories of readers who have been away from these issues for a while and to familiarize general readers not intimately acquainted with the details of the subject. Others more familiar may give this a cursory glance, if only to see where we are headed in later chapters, as this overview of these particular aspects will later form the central themes of the book. Our hope here is to leave readers with a vivid sense of the staggering complexity of evolution. Since one of our central contentions is that our modern understanding has saturated our assessment of the *Origin of Species* and of Darwin's achievements therein, this chapter will be a first step toward elucidating the contrast between evolutionary theory then and now and will serve to lay out specific bones of contention to be taken up in subsequent chapters.

Unity versus Disunity

Life is characterized by a strange and fascinating paradox. Looked at from one perspective, one can't help but notice its profound unity. The fundamental building blocks of all currently living forms are identical; all are built out of a basic, four-part molecular "code" consisting of four substances (proteins, to be precise): adenine, thymine, cytosine, and guanine, which combine in different ways as a kind of genetic alphabet, of which many organisms can be made. Better known as *DNA*, these form the biological blueprint for all living things: kingdoms as far apart as animals, plants, and fungi, for instance, share the same molecular code. At the same time, seen from another perspective, life is profoundly diverse. No fewer than several million species roam the Earth today, all distinct in appearance, each with its own "lifestyle." Countless others have come and gone since life's beginnings. How can this be explained? Reconciling these two sharply opposed truths is where evolution's complexities begin, and making sense of it all has dominated the research agenda of scholars since evolutionism's beginnings in the nineteenth century.

Living organisms are organized around biological levels. For our purposes, the cellular level makes for a good place to start, since cells are in some sense the basic building blocks of living things; an organism is, in effect, a collection of cells but is not merely so. Inside cells is housed the genetic material that provides instructions for the structure, maintenance, and reproduction of living organisms. These bits of material, commonly referred to as *genes*, are built out of the four-letter molecular code (DNA) mentioned above. Thus far we have described three biological levels: the four-letter molecular code constituting the basic material for the genes, the genes proper, and the cells housing the genes. Moving upward, we can further distinguish between two biological levels that are readily observable and more familiar: the body's individual parts and the organism as a whole. While the genes of a single organism are the same throughout all the cells of that organism, the cells themselves come in a range of types serving specialized roles and functions in the body. For instance, skin cells, muscle cells, blood cells, bone cells, and liver cells are all different from one another. These distinct and specialized organs or structures combine to form a complete and functional organism, be it a starfish, an ant, or a human being.

This simple sketch of biological levels should suffice for present purposes, but it is worth saying a brief word about how these levels are linked to each other. Although all life shares the same four-letter molecular code, the vast, nearly infinite number of ways of combining the four letters means that each species possesses its own unique set. This is reflected in the sum of genes shared by all the members of the same species. This larger set of genes shared by a species is called a *gene pool*. Similarly, the number of possible gene combinations between members of a sexually reproducing species is so immense that only a miniscule portion of that variability can be expressed in a living form at any moment of its evolutionary history. Each living creature thus represents, in a sense, a single throw of the dice. The picture that emerges is one of organisms whose separate biological levels are interconnected in complex ways. Indeed, while the gene pool of a particular species is based on the four-letter molecular code, that code is open enough to allow for the rise of a myriad of distinct gene pools, which is reflected in the number of existing species. Similarly, while the gene pool of a particular species is responsible for the range of variations observable among its members, a significant part of that genetic variability is never expressed. Again, the astronomically large number of possible combinations far outstrips the capacity of any species to actualize them through reproduction in a world where resources are always limited.

The organization of species around biological levels is usefully captured in the fundamental distinction between *genotype* and *phenotype*. The *genotype* refers to those parts of an organism unobservable to the naked eye, such as the genes and their constituent molecules. These can be thought of as the "lower" levels of living matter. The *phenotype* designates the parts observable to us, the structures and organs that make up a whole organism, the "higher" levels of the biological organization of living matter. This distinction enables us to reframe our earlier question about the unity of life as, "how can we reconcile the evident unity (uniformity) of life at lower levels with its disunity (apparent diversity) at higher ones"? This leads us to wonder whether the aforementioned unity is always explained by genealogical links between living forms, and the disunity by their independent origins, a question that might appear readily answerable in the affirmative. Unfortunately, evolution's complexities do not allow for such a straightforward reply. To see why, we must first look more closely at the lower biological levels and the several issues surrounding them.

Monophyletism and Polyphyletism

We know from the fossil record that around 3.5 billion years ago, simple, single-celled organisms inhabited the surface of Earth. These sprang out of a transformation process called "abiogenesis," which gave rise to basic organic compounds, the building blocks upon which life rests. The gradual cooling of the Earth from its initial molten state created the physical conditions for the rise of more complex organic molecules and self-replicating units (such as DNA), which have since greatly multiplied and evolved into an impressive array of simple and complex life forms.

What is not entirely clear is whether primitive life arose only once or several times during these ancient geological times. Two possible scenarios are in play. According to the first (see Figure 1), life is thought to be *polyphyletic* (the corresponding view is known as *polyphyletism*), having several ancestral sources with life arising independently in different geographical locations and/or different geological horizons. Conversely, one might also suppose life is *monophyletic* in its origins, that it shares a single common ancestor. The fact that the wide diversity of life forms springs from the same DNA strongly favors

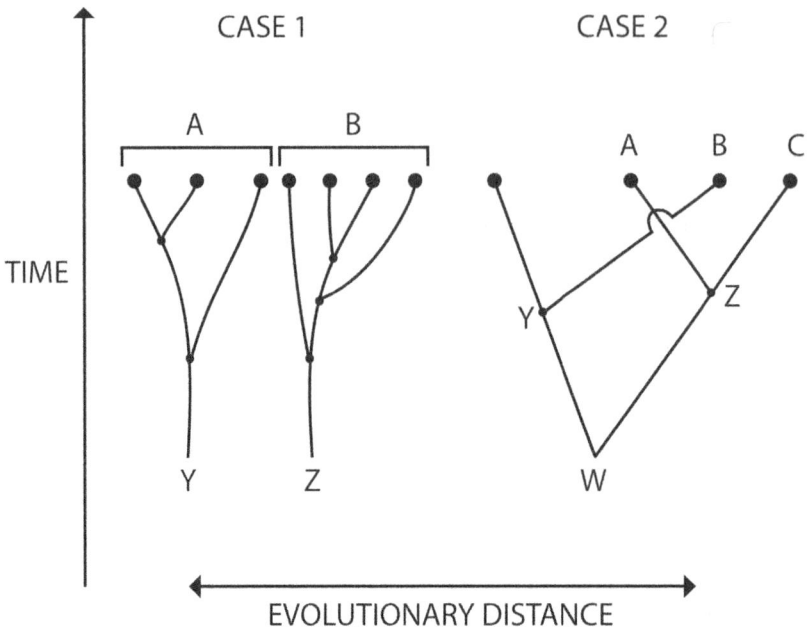

Figure 1 Polyphyletism. Evolution is *polyphyletic* when a group of life forms has several ancestral sources. Two examples are provided here. In case 1, all A and B forms when taken together are said to be polyphyletic since they independently arose from two distinct sources, "y" and "z," respectively. In case 2, although forms A, B, and C are ultimately bound in common ancestry (as seen in "w"), the fact remains that in a less distant past they arose from two separate and immediate ancestors, with B deriving from "y" and A and C stemming from "z." With regard to these more recent ancestors, A, B, and C are said to be polyphyletic when taken together.

monophyletism, the hypothesis that there is a single and unique ancestry to all life, at least on Earth (see Figure 2). Of course, this does not exclude the possibility that life on our planet went through a number of false starts in ancient times, with primitive life forms independently appearing here and there and maintaining themselves for various amounts of time, each finding different pathways and bridges across the nonliving/living threshold. Since only DNA-bearing organisms now remain, polyphyletism would imply that all but one genealogical strain of life—the one carrying DNA—has managed to survive and give rise to all the forms alive today. This explanation assumes the monophyletic origin of DNA carriers, while at the same time claiming that all other non-DNA carrying strains met with extinction.

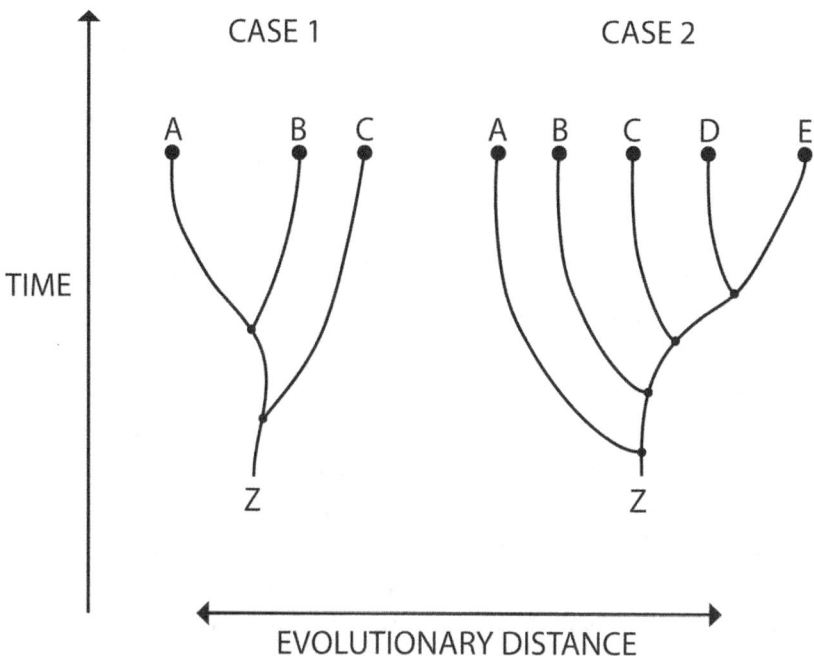

Figure 2 Monophyletism. Evolution is *monophyletic* when a group of life forms has a single ancestral source. In the two cases shown here, all forms share a single and unique ancestral root in the common ancestor "z."

Despite appearing at first blush somewhat far-fetched, another possibility should be considered here, if only to illustrate the staggering complexities of biological evolution. Framing it as a question, we might ask whether several "alternative" DNA strains might have arisen independently of each other on ancient Earth, hitting upon precisely the same "solution." While this is surely *possible*, it also runs against a deeply held assumption that there is but one sustainable way for living creatures to arise out of inorganic materials. If this were the case, it would mean that what we take today to be the unity of life bound together by monophyletism under a single DNA code is, in reality, hiding the polyphyletic origin of DNA and life in general. While this theoretical possibility should be contemplated in the name of a fully open and investigative science, it seems unlikely on its face, since it would directly imply that a complex macromolecule like DNA—whose function, again, is to encode the instructions for the structure, maintenance, and reproduction of living organisms—had been invented more than once on Earth. If one accepts that the more complex something is, the less likely it is to be replicated, this hypothesis would appear to have little going for it.

However, while polyphyletic DNA seems a longshot, it would be a mistake to reject polyphyletism as we climb to the level of the genotype, since we cannot underestimate the possibility of independent original sources for *phenotypes*. Of course, we are all familiar with look-alikes among people who appear so similar to one another they might almost be twins. Were they twins, however, this phenotypic similarity would be easy to account for: since they share the same parents, they also share the same genetic material. Look-alikes, who do not share parents, are less-than-perfect copies of each other. Nonetheless, the fact that humans share a common gene pool means that even remotely related individuals can sometimes look quite similar. Close similarities between unrelated creatures at the phenotypic level also testify to the vast arsenal of possibilities at nature's disposal, its dynamic power to create what seem like "carbon copies" by different routes, highlighting, the contingency of any existing order.

Tackling the issue of polyphyletism at the phenotypic levels requires moving beyond the confines of a single species to a consideration of how life in general interacts with the outer world. Since its formation some 4.5 billion years ago, Earth has experienced continual change. We can think of these changes as falling into two types. Some emerge from the intermittent but perpetual geological events on our ever-changing Earth, which can subtly or radically change the playing field on which organisms must compete. Examples of such events include the gradually cooling of the Earth from its original molten state, the migration of continental lands on its surface, and episodes of glaciation or ice ages and their subsequent retreat. The list is long. We are now confronted with an all-too-apparent example in the form of global climate change: the Earth's warming—albeit due to human factors—has made previously hospitable environments hostile to some organisms and vice versa: one can think here of changing growing seasons or the highly sensationalized northward march of fire ants, killer bees and, more recently, so-called Murder Wasps. These environmental changes are called *abiotic* factors. A second type of change is brought on by the encounters between life forms. A life form rarely has free reign to exercise its capacities to exploit nature's resources. Earth has been inhabited by a myriad of forms for a very long time, all of which need resources to live. During the first few billion years of Earth's history, single-celled organisms were the dominant form of life, with much more complex forms first appearing 500 or 600 million years ago. Earth has been pretty crowded ever since, posing many challenges to each species, who must share whatever food and accommodations a given habitat provides. Transformations brought about by the interaction between life forms are known as *biotic* factors.

The combination of *biotic* and *abiotic* factors exposes life to permanent challenges. If evolutionary change itself is a central reality of life, it is precisely because evolution is often the only way to respond to these challenges: to put it bluntly, the imperative is often as simple as "change or die." This response is called *adaptation* and is most conspicuously reflected in the phenotypes of species, each differently adapted to make their living. Ever faced with the risk of complete and definitive extinction, life forms try to find ways to adapt to new conditions. Evolution has quite a bag of tricks at its disposal to help life forms respond to these constant and ever-changing challenges. In order to see why polyphyletism at higher biological levels is still a live option in explanatory terms, we will review some of the most prominent tactics nature employs, which form well-worn and recognizable pathways as it rolls out through time.

Anagenesis or Continuous Evolution

Just as you look (more or less) like your parents, you also differ from them somewhat, of course. When they made you, your parents transmitted the genes that made them what they are, which explains many of your shared physical similarities. At the same time, you are no carbon copy: differences were also introduced, a factor inherent in the very nature of sexual reproduction. The combination of your parents' genes during your conception introduced some novelties reflected in, and contributing to, your uniqueness, your individuality as a physical being. Your sisters and brothers, should you have any, are not exactly like you, since your parents handed down different genes during each separate act of conception. It just so happens that sexually reproducing organisms contain a significant reservoir of genetic variations that can only express themselves at conception time. The only exceptions to this are identical twins. In this case, exactly the same genetic material is used for the conception of two individuals or more. Of course, twins are fairly rare.

Applying our reasoning to the entire human species, we can see how even in the simple process of reproduction the dice are cast and so automatically introduce, with each generation, novelties or new variations. The introduction of randomness in sexually reproducing species also endows a protective mechanism; continually generating new sets of variations helps such species cope with new external conditions through adaptation. This periodic injection of randomness acts as a mechanism for generating new evolutionary solutions, raw material for tomorrow's life-saving adaptations.

One important piece is still missing from our understanding of the multiple and independent (polyphyletic) rise of phenotypic features. The existence of biological variations is merely a precondition for adaptation. Something more is needed to set the process of adaptation in motion: an encounter between the potential variations available to a specific species *and* the external conditions (biotic and abiotic) at a particular place and time in evolutionary history. If the climate becomes colder over a given stretch of time, for instance, organisms belonging the same species with resistance to the cold are more likely to survive than their conspecifics not so equipped. Through the grace of genetic variation, they may have wound up with, for example, a more efficient thermoregulatory system for internal heat, a thicker skin, or more body hair. The features that give such organisms an edge in a cold climate are precisely those that help them live long enough to reproduce and perpetuate the species in the next generation. Conversely, organisms unable to cope successfully with the colder climate will die out, no longer contributing to the set of variations transmitted to future generations, in effect withdrawing their variations from further consideration. The simple differential process that consists in selecting the best-fit organisms and excluding others within a species is what *adaptation* is all about. One can imagine that in some instances, having a large body might be an advantage, in others, a potentially fatal disadvantage. Think of a situation in which a number of species inhabit the same territory, thus competing for the same resources. One way to outcompete other species, in this context, might be through the selection of individuals with increased body size: bulky species can gain an edge by intimidating smaller ones or controlling them in various other ways. When, say, a bear or elephant appears on the scene, its hulking size sends a clear message to creatures of lesser stature, who often clear out immediately, greatly enlarging opportunities. Conversely, larger size can also be a minus, relative smallness a plus. Here one needs only think of insects, the greatest of all evolutionary success stories perhaps, in terms both of the number of individuals and of species.

These examples show the ways we can think of evolution as a process that is driven and oriented in a certain direction. Species in the stream of geological time—let us call them *lineages*—can increase their chances of persisting by many means: developing fur as a means to withstand cold climate; moving more swiftly with greater agility than other species in their neighborhood; being so small as to inhabit places other species cannot occupy; becoming more aware through the senses (smell, hearing, sight); or by being more intelligent than others. There is no limit to this list of strategies and the advantages they provide, which is in principle open-ended. Of course, "advantage" is defined by context and no given trait is valuable in itself, even intelligence. Humans are an obvious case in point: think of the countless changes to planet Earth wrought by humankind unintentionally through its "intelligent" activity. It is at this point an open question whether intelligence will provide the means for correcting its current course, or whether our species will be just another failed experiment, like millions of others. A compelling metaphor for the mixed blessing of human intelligence as an evolutionary trait can be seen in the Great Pacific Garbage Patch, one of several miasmas of degrading plastic, chemical sludge, and wood pulp (likely originating from toilet paper), which at this writing has begun to rival some significant land forms in size (current estimates rank its size somewhere between Texas and Russia). Interestingly (but unsurprisingly), new creatures have begun to evolve specifically adapted to this largely "artificial" plastic habitat. Climate change is only the most obvious threat that can be chalked up to human intelligence; others loom.

Which features are developed and which are not entirely depends on what variations are available throughout the evolutionary history of each lineage. Species cannot modify themselves at will but must instead wait for chance to throw the variations their way at just the right time. This really brings home the value of continual injections of random variations: to persevere, a species must be lucky enough to have among its members the right biological variations needed for survival at the right place and time. Again, variation is a necessary but not sufficient precondition for evolution. It is the selection of the available variations over generations that is responsible for evolutionary modification (and adaptation). When the history of life on Earth is viewed broadly, we see that the wide array of distinct evolutionary lineages—such as mammals, birds, reptiles, fish, insects, mollusks, corals, and worms—have taken off in all sorts of evolutionary directions. The overall picture is a welter of unique and separate routes that defies the imagination.

Thus far, we have been considering the selection process of variations available within a single lineage that lead to adaptations in one particular direction. This is called *anagenesis* (see Figure 3). Under anagenesis, the transformation of a lineage over geological time is an unbroken, continuous process, meaning that the members of each generation of a species continue to share in the rise of new features through reproduction, thus maintaining the unity of this evolving species or lineage over time. The notion of *anagenesis* is more easily understood when considered alongside other evolutionary possibilities, a brief review

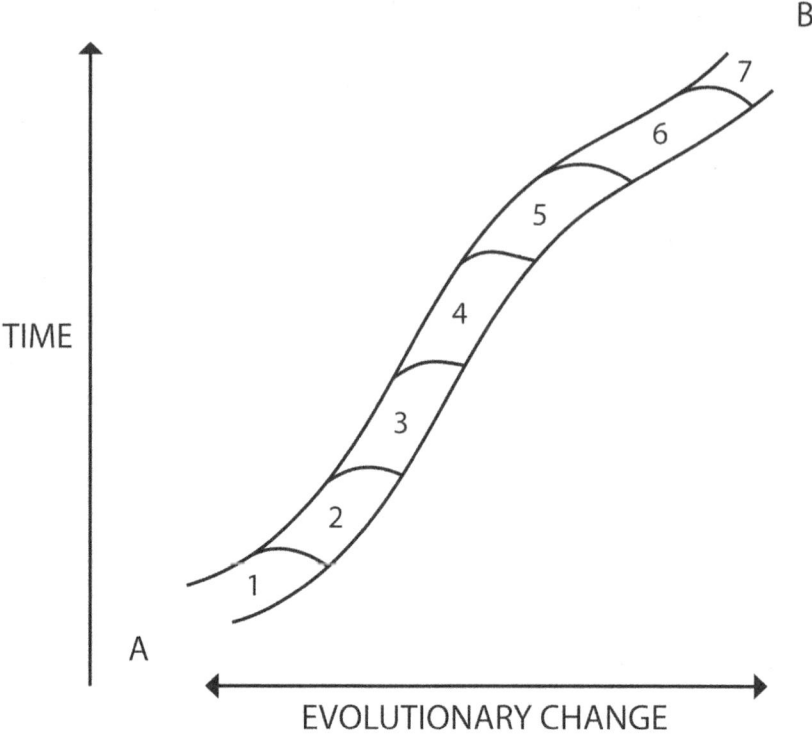

Figure 3 Anagenesis. This is the name given to the transformation of a lineage in an unbroken and continuous fashion over geological time, so that members of each generation of a species (in our example seven separate generations) continue to share in the rise of new features through reproduction without breaking apart, thus maintaining the unity of this evolving species or lineage over time. In our example, even though A and B may be different enough to require two different appellations, the single common genealogical nexus to which both belong remains uninterrupted.

of which we present below. As we shall see, these evolutionary processes and patterns available to any given species, and traceable through their lineages, add yet another layer of complexity and variability to evolution.

Cladogenesis and Divergence

With more than several million distinct species alive today, evolution must have had at its disposal a process for generating new lineages. In other words, the history of life concerns not only the evolution of a single species or lineage over time, as discussed above; it also involves the proliferation of the number of species themselves. Brand-new species are generated by subdividing or splitting off, a process called *cladogenesis* or *speciation* (see Figure 4). This is explained by the following: organisms belonging to the same species are sometimes confronted with different pressures to adapt across the species range, pulling that species apart. Indeed, the more populous a given species becomes, the greater its geographic footprint, meaning it is more likely to face novel conditions as it extends its geographical range. These conditions may vary significantly; the climate may be warmer or cooler, for example, at more remote edges of the territory; and the roster of competitors may differ. Breaking new ground makes for new challenges and opportunities.

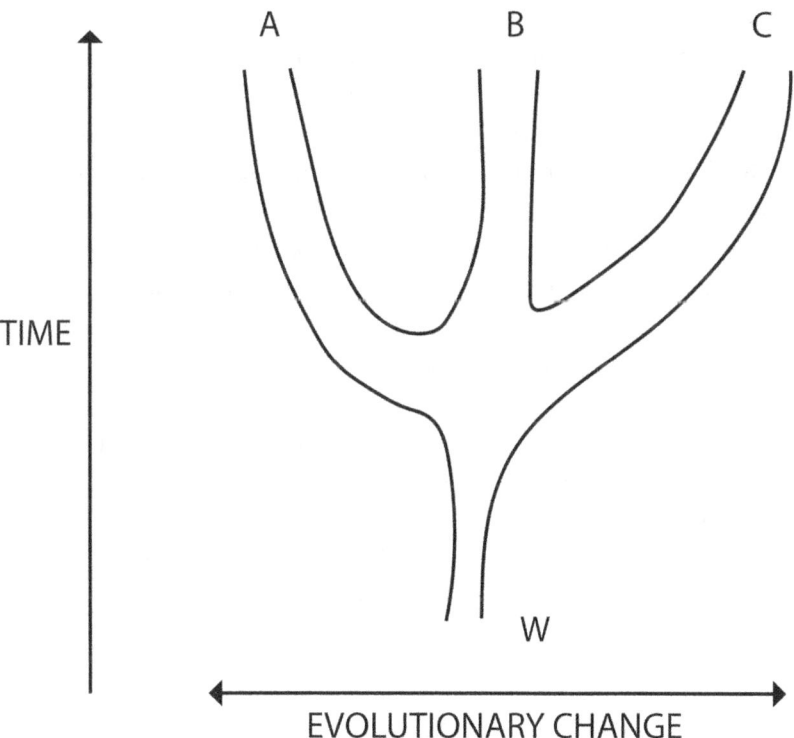

Figure 4 Cladogenesis. This name refers to situations in which new species are generated by the subdivision of a genealogical nexus. In our example, three new species (A, B, and C) are born following the splitting-off of a single common ancestor "w." In this book, *cladogenesis* is used synonymously with *speciation*.

Irrespective of the fact that organisms belonging to the same species are connected through reproduction, the pressure to find different adaptive solutions across the geographical range may eventually lead to the break-up of that species. While the continual flow of gene exchanges during reproduction tends to keep a species united, the relative isolation of several of its groups of organisms—as when they cross geographical barriers like mountain ranges or large rivers, for instance—creates ideal conditions for the eventual break-up of that species when these physical barriers prevent reproduction. Out of the original species (the "parent species") may issue in two or more distinct species (the "daughter species") following its break-up. Furthermore, *cladogenesis*, or speciation, is often accompanied by some degree of evolutionary *divergence*, if only because the several daughter species that have arisen out of a single original parent species now simultaneously evolve away from each other, each having found its own unique adaptive solution to the different challenges encountered.

There is more to the notion of *divergence*, however. Two life forms not closely affiliated may still be said to be diverging from one another when some of their features are compared. For instance, two middle-sized forms are said to be diverging if one of them increases in size over geological time while the other decreases in size over the same time period. Depending on the set of features observed, the forms being compared may diverge as just explained, or keep the same distance relative to one another (this is called *parallel evolution* or *parallelism*), or move even closer to each other with features growing more similar over time (see *convergence* below). Under the continual pressure for adaptation resulting from biotic and abiotic challenges, lineages evolve in all directions through a combination of the aforementioned patterns of divergence, parallelism, and convergence. (To look ahead briefly, we will see in later chapters that Darwin is quite a fan of "divergence" and grants it pride of place in his theory; this will end up having significant consequences for the ultimate contours of his theory and militates against it being considered a genuinely *modern* theory.)

Reticulate Evolution

When *convergence* occurs among very closely related forms, the pattern can lead to *reticulate evolution* (see Figure 5), a kind of failed cladogenesis or speciation. The process involving a parent species on the verge of splitting into two daughter species is interrupted when the latter two forms eventually manage to

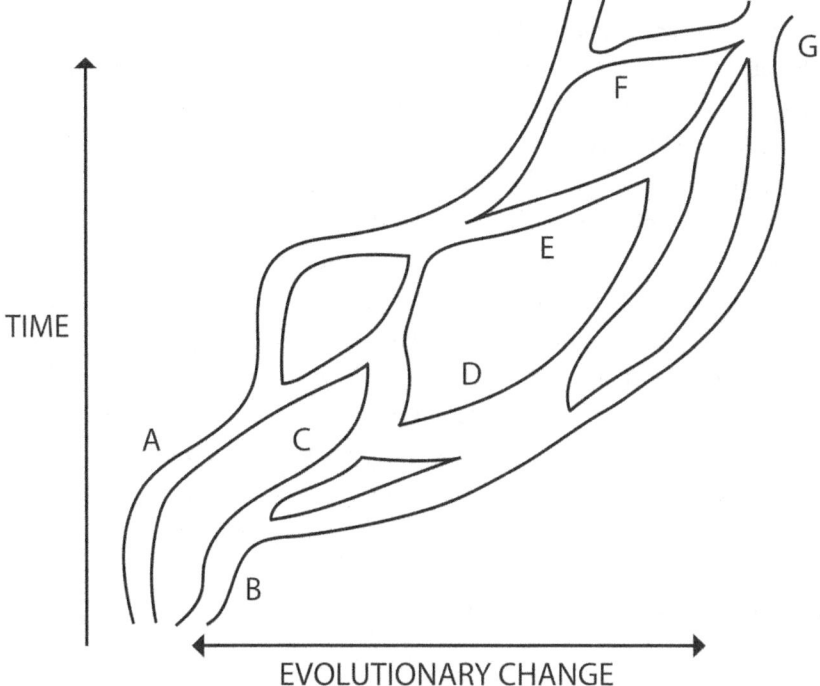

Figure 5 Reticulate evolution. This process involves a group of forms or strains (from A to G) on the verge of splitting into several distinct forms but is interrupted when these strains eventually manage to re-establish reproductive contact with one another or the main genealogical nexus. This is a kind of failed cladogenesis or speciation, an ongoing fission-fusion process generating a complex network of very closely related forms bound in a single genealogical nexus.

re-establish reproductive contact with each other. This can be the case, for instance, when two closely related forms find a way around geographical obstacles—for example, mountain ranges or large rivers—that had previously obstructed their free movement and contact with each other or when a climate shift decreases the adaptive pressure, that is, when the driving force behind their original split weakens or disappears.

Under such circumstances, one can imagine the evolution of a group of closely related forms interconnected in a complex reproductive network. While forms belonging to that network periodically diverge away from others, most of the time they fail to reach the irreversible break-up point where interbreeding is impossible. Extended over geological time, reticulate evolution offers an evolutionary pattern of closely related forms continuously diverging and converging in an ebb and flow of partial breaks followed by reunifications. This pattern has been likened to an ongoing fission-fusion process; forms splitting away are often reabsorbed within the main reproductive stream.

Equipped with this background knowledge regarding *reticulate evolution* and *divergence*, let us revisit the question of *monophyletism* versus *polyphyletism*. When two daughter species manage to irreversibly move away from a common parent species in separate evolutionary directions, the latter two are said to be bound under monophyletism through their sharing of a single and unique common ancestor. In this case, many similarities shared by the two daughter species are explained by a common heritage. However, the question is blurred when one is dealing with reticulate evolution. When a group of closely related forms is implicated in a complex reproductive network characterized by a fission-fusion process of incomplete splittings as discussed above, it is harder to say in retrospect whether similarities between two particular forms, for instance, should be attributed to a single ancestor or to several of them. By definition, reticulate evolution involves a complex reproductive network composed of several forms. Looking at such forms, one can't rule out a polyphyletic explanation for their similarities and differences.

Convergence and Analogies

Convergence, too, has other significant features (see Figure 6). If one had to boil evolution down to just two aspects, one could explain similarities between forms either by kinship (common parents) or by modification over time (adaptation). Had we lived in a simpler world, similarities could always be chalked up to genealogical (or phylogenetic) connections, dissimilarities to adaptation. As the reader may have already surmised, matters here, as elsewhere in our story, are not so simple. The evolutionary engine mixes up, blurs, and obscures the genealogical (phylogenetic) traces as it goes forward. True, similarities are often explained by common ancestry, but it is not uncommon to encounter similarities generated not by a common parentage but by evolution having found similar solutions to the same problems or challenges.

For instance, it is no accident that many aquatic and marine creatures have roughly similar torpedo-shaped bodies; one might think here of the similarity in appearance between sharks and dolphins. Genealogically speaking, these two types are very distant from each other—having branched off from a presumed common ancestor perhaps as far back as 450 million years ago—and belong to two distinct classes of animals, the former being a cartilaginous fish and the latter a mammal. Furthermore, while sharks have always lived in water, dolphins

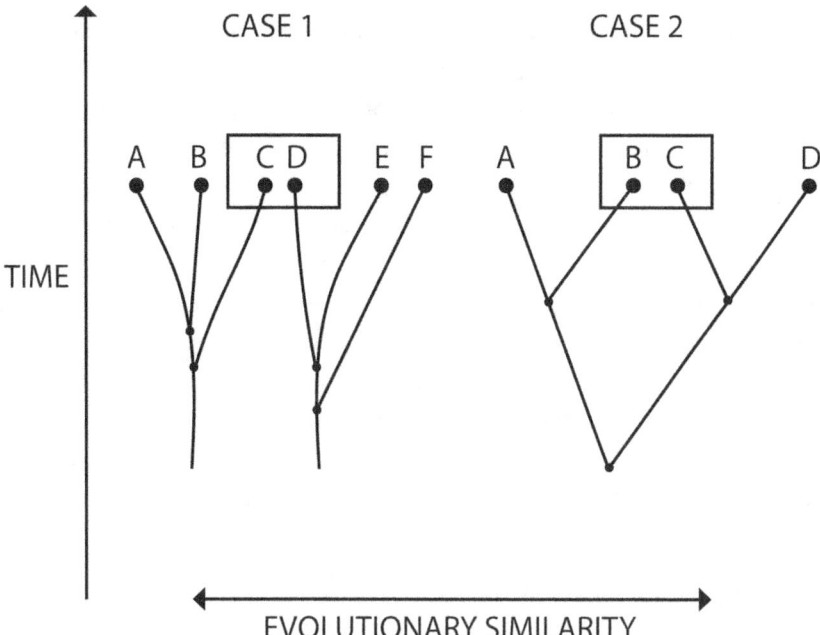

Figure 6 Convergence. While similarities between forms are often explained by the sharing of a common ancestor, it is nonetheless not uncommon to see the rise of similarities generated by evolution having found similar solutions to the same problems or challenges. When this is the case, natural selection forges similarities (also called "analogies") between forms that are said to be *converging toward* each other. In case 1, we see two forms (C and D) not closely related evolving closer from each other. In case 2, two forms (B and C) fairly closely related also converge toward each other.

evolved out of terrestrial ancestors that lived some 50 million years ago. Can we explain their similarities by the fact that both are chordates? Considering that humans and birds are chordates too, one can easily see similarities between a shark and a dolphin are not well explained by genealogy or phylogeny. The answer must be sought in their common habitat: the oceans. Because water is a medium much thicker than air, evolution shaped many elongated "fish-like" creatures to move more easily through it, with all the advantages this entails (including lower energy consumption and faster flight from predators, or approaches toward prey).

The similarities between sharks and dolphins are, therefore, best explained by *convergence*: similar adaptations resulting from similar challenges to life. These are called *analogies*, that is, similarities explained by adaptation and not by genealogy. Of course, by looking closely at the anatomy of sharks and dolphins, their differences become more obvious. After all, the two forms are so remote, phylogenetically speaking, that their differences are manifold. But what if one is dealing with forms more closely related, such as the Philippine flying lemur and the flying squirrel—bound in common ancestry perhaps some 90 million years ago—two small mammals adapted for gliding and belonging to two distinct orders (Dermoptera and Rodentia, respectively)? In this case, it becomes more difficult to distinguish between similar features explained by common ancestry (*homologies*), and *analogies* or *analogical* features attributable to their adaptation to the same lifestyle (in this case, gliding through the air). The question of whether we are looking at *homologies* or *analogies* can easily become blurred in the context of life forms not too remotely related and, therefore, whether similarities between two forms are explained by a single ancestral source (*monophyletism*) or by several (*polyphyletism*).

Phylogeny versus Adaptation

At this point we should also return to our prior discussion of the opposition between *phylogeny* (genealogy) and *adaptation*, as this will allow us to penetrate more deeply into the intricacies of evolution. Let us raise a simple question concerning *phylogeny*: why should descendants of a given lineage retain a set of features over time? The answer is their *adaptive value*: if some ancestors have managed to survive, they must have gotten something right. Their constitution fit circumstances well enough to help them carry the day. These successful features, unsurprisingly, were then transmitted through reproduction to subsequent generations, who also passed on their useful traits with adaptive value. Phylogeny or ancestry, therefore, constitutes a *conservative* aspect of evolution: "good" features are preserved and transmitted over the course of geological time. Because of their adaptive value, what were once innovations, that is, new and useful adaptations, become stable features (defining characteristics, one might say) passed along in the phylogenetic stream. New, essentially random features eventually become stable features to be passed along to future generations.

However, the master of the game in evolution is not its conservative aspect represented by *phylogeny*, which preserves old adaptations, but rather its wild, creative side: the random changes that come to be new *adaptation*s that give life an edge in confronting immediate and challenging conditions. There would be no reason for life to change if the milieu always remained the same. Were this the case, well-adapted forms of years past would merely transmit precisely those features that have always worked before. To reverse the perspective, there would have been no need for evolution to devise life forms endowed with internal mechanisms for generating variations and novelties. Because the milieu keeps changing, species are condemned, and enabled, to seek out new solutions. We have seen that species carry a reservoir of genetic variations to help them respond to such changes. Yet, despite this built-in protective device, 99 percent of all species that have ever existed are now *extinct*. The fact is that nearly all species have a limited lease on Earth.

This fact powerfully illustrates the harsh reality that life on Earth is often confronted with the need to evolve, *adaptation* or *innovation* being the way to improve the odds of sticking around. When they fail, older species are replaced by others better adapted to new conditions—new species that arise either from the process of splitting (*cladogenesis*) or from the process of profound and continued change over time within the confines of the same lineage (*anagenesis*). Indeed, when lineages are modified to a considerable extent, it is no longer possible to consider them members of the same species; new names and categories must be created for them (a matter we will return to in Chapter 2).

The Pace of Evolution

The pressure on life created by the surrounding milieu varies dramatically in terms of both time and place and is continually transforming itself. This means the pace of biological evolution can vary greatly, ranging from near stagnation to rapid transformation, and is constantly shifting gear. Changes to the milieu can be local or global, impacting a smaller or greater number of lineages. For instance, around 65 million years ago, Earth was devastated by an episode of mass extinction killing roughly 75 percent of all species—one of several such mass extinctions we are now aware of—resulting in the extinction of major groups of organisms such as the dinosaurs (among the reptiles) and the ammonites (among the mollusks). It is believed that a global environmental change following a comet or asteroid impact disrupted the whole food chain on Earth. Considering that life is always opportunistic, the void created by this cascade of extinctions opened up new possibilities for the surviving forms, including the mammals (their loss was our gain, so to speak). Suddenly, evolution accelerated through the proliferation and modification of previously existing forms and the rise of new ones such as the whales, the primates and the horses, among many others, some 65 to 45 million years ago.

The Genotype and the Phenotype Revisited

We have characterized *phylogeny* (genealogy) as the more conservative factor in evolution, passing on tried-and-true features to future generations, with new *adaptations* best conceived as the innovative and disruptive components driving evolution forward in new directions. At this point, we can already see many factors in play. Matters become even more mind-boggling when we take into account our earlier discussion concerning the distinct biological levels found within organisms and species.

At the most fundamental level of the *genotype*, all currently living forms are DNA carriers constructed around the same four-letter molecular code (adenine, thymine, cytosine, and guanine). At this level, life leans heavily toward the conservative, having retained the same unique and traceable connections binding all life forms into a single genealogical or phylogenetic stream, as postulated in monophyletism. Interestingly, the ever-pressing need for life to evolve and adapt has not severed this common and conservative thread uniting them all.

Things look rather different, however, as we climb up to the highest biological levels of the *phenotype*, that is, as we move from the four-letter molecular code to the genes proper, to the cells, to the tissues, to the organs, and to the organism as a whole. Although all levels are interrelated in complex ways within the same organism, life forms are nevertheless more directly confronted with the world's harsh realities at their higher biological levels. It is with their phenotype that organisms engage the outer world, that is, with what can readily be observed of them: body size and shape, number of arms and legs, and kinds of sensory organs. The phenotypic features experience the greatest modification through the course of evolution in such a way as to provide new adaptive solutions to biotic and abiotic challenges.

To put the matter somewhat differently: the organism is a unified whole, with genotype and phenotype being simply two sides of the same coin. The genes carried by an organism and the observable features of that same organism are somehow connected. After all, phenotypic features are gene based. However, the relationship between the biological levels constituting the same living organism is anything but straightforward, with complex feedback loops or interactions between the levels. While the specifics are beyond the scope of this book (see "Annotated Bibliography" for more detail), it is enough for present purposes to acknowledge that deep down inside living matter one finds strong proof of the monophyletic origin of life, including the universality of the four-letter molecular code of DNA carriers mentioned above. At these lower levels, life is somewhat sheltered from adaptive pressures. This is not the case at higher levels. There, the pressure to adapt to biotic and abiotic changes has instigated profound modifications in the phenotypic features over geological time. Under these conditions, similar evolutionary solutions to similar challenges were often found by life forms not necessarily very closely related. When this is the case, similarities become mere *analogies* and not *homologies*, rendering obscure the issue of what is to be explained by a common origin (monophyletism) or by several distinct and independent origins (polyphyletism).

An illustrative example of this difficulty can be seen in the advent of an evolutionary novelty like the eye. The novel adaptive benefits of having eyes in the struggle for survival are readily apparent, whether this comes from enhancing competition against other species or from exploiting natural resources more effectively. Although one might presume that such a complex organ as the eye could have arisen only rarely during evolution—favoring a monophyletic explanation—we know from closer analyses that eyes arose many times independently of each other in a wide and varied range of shapes, sizes, and designs. The adaptive advantage of developing eyes was so great that evolution somehow found the way to invent them over and over again, independently of kinship. In the case of eyes, therefore, polyphyletism cannot be so easily discarded—in fact, it is the explanation we now know to be correct in many cases.

An exhaustive list of evolution's complexities (and their underlying explanations) lies beyond the scope of the current work, but those considered above will provide sufficient background as we begin our discussion of Charles Darwin. As these notions are a crucial part of our narrative, we will have occasion to return to them throughout the book. Ultimately, we aim to show that, while Darwin deserves our profound gratitude for bringing some of these complexities to light, there are others he either glosses over, downplays, or even ignores.

Conclusion

The modern view of evolution is based on the realization that the process has been complex since its very beginnings. While DNA carriers are somehow united in a single giant genealogical network, time has to a greater or lesser degree erased the exact links connecting many of them at the phenotypic level. Under the adaptive pressure to confront biotic and abiotic challenges, life was forced into an "arms race" in the search for new evolutionary strategies, leaving an impressive number of extinct forms in its wake. The overall picture of evolution

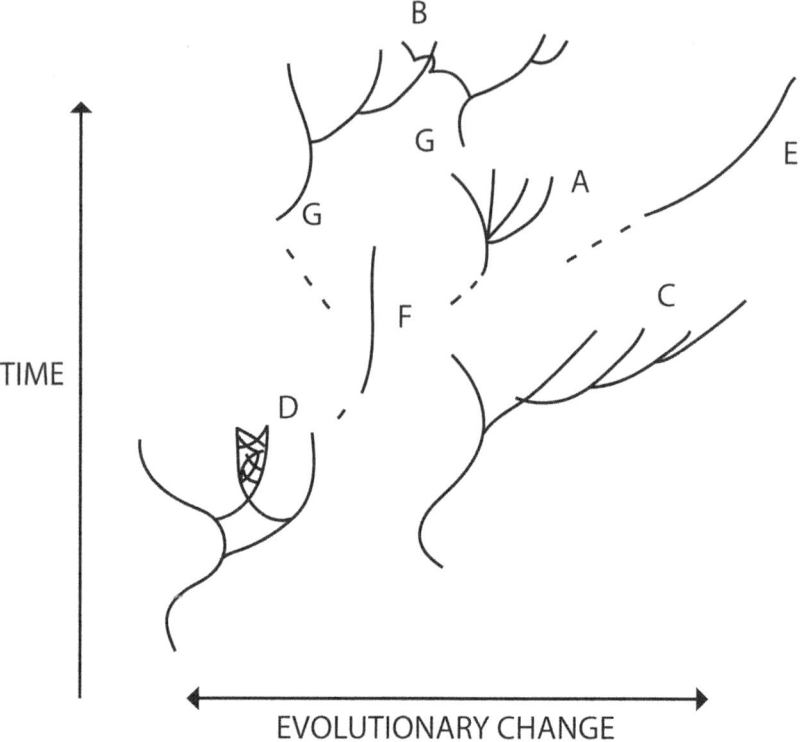

Figure 7 Random evolutionary walk. A truly open evolutionary process is expected to generate the full range of evolutionary patterns over time, with many forms being lost (extinction) in the depth of time, finding no living counterparts and having no obvious phylogenetic connections between them. Multiple manifestations should be reflected under: (a) divergence and monophyletism, (b) convergence, (c) parallel evolution, (d) reticulate evolution, (e) anagenesis, (f) stagnation, and (g) polyphyletism. For his part, Charles Darwin chose to organize evolution mainly around a dominant pattern, "divergence," hoping it would offer him a window into the most important evolutionary events of the past. By making that assumption, Darwin closed off any possibility of recognizing the full complexity of evolution (compare with Figure 9).

that emerges is that of a "random evolutionary walk" (see Figure 7), with some lineages moving away from each other (divergence), others maintaining the same distance relative to one another (parallelism), still others moving closer to each other by developing similarities or analogies (convergence), with a number of other lineages being engaged in a fission-fusion process (reticulate evolution). In addition, the pace of evolution among separate and isolated lineages has continually slowed or quickened according to localized changes of the milieu. At times, whole groups of lineages were simultaneously and profoundly impacted by global environmental fluctuations leading to events of mass extinctions, opening up new opportunities for forms that survived these catastrophes. Modern evolutionary theory has fully embraced the knotted intricacies of the monophyletism-polyphyletism spectrum and is keenly aware of the difficulties involved in disentangling them.

Annotated Bibliography

A great number of competing theories of evolution have been proposed since the nineteenth century, each trying to make sense of evolution's complexities to various degrees. Because this book focuses on achieving a better understanding of Charles Darwin's *Origin of Species*, a formal exposition of such a wide range of theories is beyond our present scope. In this chapter, we have hoped to provide readers with enough detail to prepare the way for our analysis of Darwin's work. With this in mind, we thought it useful to provide a more-detailed annotated bibliography for readers interested in exploring how evolution's complexities might be explained under competing theories and hence provide a sketch of some contemporary issues relevant to the foregoing discussion, as well as to provide resources for further investigation.

Let us begin by revisiting the issue of the different biological levels. While higher biological levels (organs, structures, whole organisms) rest upon lower biological levels (molecules, genes) for their existence, the phenotype is not rigidly determined by the genotype. A certain amount of indeterminacy is thus introduced between the biological levels. In other words, while observable features require a genetic basis, genes themselves and their coding pathways (i.e., proteins) may not necessarily find expression at the observable level. There are several reasons for this. As stated in this chapter, the gene pool of a particular species contains more genetic variability than can possibly be expressed by its sexually reproducing members at any given moment of the evolutionary history of a lineage. Furthermore, and more importantly for our point here, the genetic makeup of a species contains a great degree of genetic information that is inactive or "noncoding," playing no role (or no known role) in the phenotype. This so-called "junk DNA" is the product of long periods of evolutionary change, accumulation, and duplication of genetic material no longer required in the elaboration of organisms. Organisms thus carry more genetic information than required for their immediate survival needs. When changes or mutations occur within the inactive genetic material, not only do these find no expression at higher levels, but they can also be used to evaluate the differences that have accumulated between different species in the intervening years since they were separated from a presumed common ancestor.

The rationale is the following: since inactive genetic material plays no immediate role in the survival of organisms (by finding no expression in the phenotype), it is free to change under *genetic drift*, for instance, independently of the other features (genetic or otherwise) that are directly confronted with the adaptive scrutiny of external conditions. This thinking gave rise to the "neutral theory of molecular evolution" developed in the 1960s and 1970s, which also served as the theoretical basis for the idea of a "molecular clock." If inactive genetic material is modified independently of the external conditions—the argument goes—such modifications can be more revealing of genealogical connections and separation time (speciation) between species than phenotypic features that have been more directly impacted by adaptation. The recent study of entire gene pools belonging to a number of species (including humans) under the rising field of "genomics" has shown that different segments of a species' gene pool evolve at varying paces, implying that the same species carries more than one molecular clock within itself. Furthermore, it is now understood that molecular clocks can be, and occasionally have been, impacted by the external world, for instance during major global environmental changes associated with mass extinction events. Through all these scientific developments, scientists realized how the biological levels within an organism and a species are complex and not rigidly connected. And things can get more complex still with feedback loops or interactions between the biological levels within the organism or between the organism and its environment. These interactions are as yet not very clearly understood. For fairly readable introductions to some of the issues mentioned above, consult Lewin, R. (1999), *Patterns in Evolution: The New Molecular View*, New York: Scientific American Library; Kratz, R. (2009), *Molecular and Cell Biology for Dummies*, Indianapolis: Wiley; Futuyma, D. J. and M. Kirkpatrick (2017), *Evolution*, 4th edn, Sunderland MA: Sinauer.

Because nothing in biology is simple, it came as no surprise when researchers discovered that the gene pool of each species is not as unique as was once believed. It was expected that the gene pools of two closely related species would share many genes in common, and this is what was found. For instance, humans and chimpanzees share roughly 98 percent of their genes in common. The remaining 2 percent accounts for all the differences, especially considering that the modification of regulatory genes has greater impact on phenotypic development than the modification of structural genes. Indeed, while regulatory genes control whole groups of genes, the action of structural genes is much more restricted. The modification of a single regulatory gene can, therefore, cascade into a series of more or less important phenotypic changes. What was not expected, however, is the discovery that many species very distantly related share whole groups of (*homeotic or Hox*) genes in common. For instance, the development of the eye in groups as far apart as insects and vertebrates—irrespective of their profound phenotypic differences—is subject to the action of similar groups of genes (*Pax6*). We had known for some time that the molecular code (DNA) was the same for all life; we now realize that some groups of genes are widespread even among very distantly related life forms. This adds to the unity of life at the genotypic level. This newly recovered unity in some genes across a wide taxonomic range is not, however, accompanied by phenotypic unity, as testified to by millions of species differently adapted to face the outer world. The loose connections between the genotype and the phenotype are thus further confirmed by these new discoveries. Consult Sundaram, V. et al. (2014), "Widespread Contribution of Transposable Elements to the Innovation of Gene Regulatory Networks," *Genome Research*, 24: 1963–76.

The distinction presented in this chapter between the genotype and the phenotype was used as an introduction to the different biological levels. In fact, the last decades have seen an ongoing, ever-more sophisticated debate about both the number and the nature of these biological levels. For instance, if genes are the only entities that are able to persist across generations through reproduction, then perhaps bodies or whole organisms carrying them should be considered as minor players in the evolutionary game. What is ultimately being selected for long-term survival—the argument goes—are only good genes and not entire organisms, whose role, on this view, is limited to the mere transmission of genes to the next generation. Adaptation should therefore be understood as a process largely focusing on genes. For this "gene-centered" view of evolution, see Dawkins, R. (1976), *The Selfish Gene*, Oxford: Oxford University Press and Dawkins, R. (1982), *The Extended Phenotype*, Oxford: Oxford University Press. The approach giving precedence to the genotype over the phenotype in evolutionary studies gained momentum among many evolutionists following the discovery of the double helix structure of the DNA macromolecule by Francis Crick and James Watson in 1953. Since that time, the field of molecular biology has become central in evolutionary biology. See Morange, M. (2017), "Molecularizing Evolutionary Biology," in R. G. Delisle (ed), *The Darwinian Tradition in Context: Research Programs in Evolutionary Biology*, 271–88, Switzerland: Springer Nature.

A less reductionistic approach has departed from the "gene-centered" view of evolution by focusing, instead, on the biological levels of organisms and species. It is argued that while "good" genes outlive their carriers (organisms), the fact remains that it is the carriers that confront the external world. It is their good phenotypic variations that are directly selected, and not their genes. The genotype is therefore largely dependent upon the phenotype for its persistence over time. Furthermore, it is argued that the species level is fundamental for the evolutionary process since it is at this level that evolutionary events play out. For instance, *cladogenesis*, or speciation is a process that involves organisms belonging to the same species. Now, whether or not this species will undergo a splitting process depends on factors such as the amount of selective pressure, the number of organisms and their variations, the structure of the species (the way it is distributed in geographical space), the kind of geographical obstacles encountered that may or may not obstruct the free reproductive flow among the species' members. A classic exposition of this kind of "organismic biology" is found in Mayr, E. (1963), *Animal Species and Evolution*, Cambridge: Belknap Press.

Whether one focuses on genes or on organisms, the kind of biology described thus far is committed to the following equation: *microevolution = macroevolution*. Whereas "microevolution" refers to evolutionary events at and below the species level (species, varieties, races, gene pools, genes, and DNA), "macroevolution" applies to evolutionary events above the species level (genera, families, orders, classes, kingdoms). The "microevolution = macroevolution" equation simply means that all evolutionary events are believed to be exclusively explained by reference to phenomena at the microevolutionary level, that is, involving entities ranging from DNA molecules to species. Differently stated, macroevolutionary events involving genera or kingdoms, for instance, require no explanations of their own, as these are merely epiphenomena ultimately explained at the microevolutionary level. Explanations of them, therefore, can only be found in microevolutionary explanations. Promoters of the "microevolution = macroevolution" equation assume that if you want to know, for instance, why and how two related forms belonging to two distinct genera or families are diverging from each other in different directions, one must simply ask how and why two distinct varieties belonging to the same species managed to split up, thus eventually becoming two new daughter species. In short, macroevolutionary events are merely microevolutionary phenomena extrapolated over geological time to higher taxonomic levels.

The "microevolution = macroevolution" equation constitutes an assumption that is common among molecular biologists, geneticists, zoologists, and taxonomists. This assumption, however, has been vigorously challenged, beginning in the 1970s, by a new generation of paleontologists—re-establishing a tradition going back to the pre-1950 period—giving rise to a field renamed "paleobiology" that is devoted to studying macroevolutionary patterns and processes. See Stanley, S. (1979), *Macroevolution: Pattern and Process*, San Francisco: W. H. Freeman; Sepkoski, D. and M. Ruse, eds (2009), *The Paleobiological Revolution*, Chicago: University of Chicago Press; Sepkoski, D. (2012), *Rereading the Fossil Record: The Growth of Paleobiology as an Evolutionary Discipline*, Chicago: University of Chicago Press. Paleobiologists hold that macroevolutionary events require explanations of their own. Just as events at the gene and species levels are different from one another—the argument goes—so there exist events unique to higher taxonomic levels (genera, families, orders, classes, kingdoms). By so arguing, some evolutionists extend the number of biological levels relevant for our understanding of evolution. Evolution is no longer seen as the gradual accumulation of microevolutionary changes extrapolated to higher taxonomic levels. On the contrary, evolution is envisioned as a hierarchical phenomenon having different levels, each under different kinds of processes that are, themselves, often causally independent of one another. If processes are different at each level in addition to being causally unconnected, microevolution and macroevolution must be uncoupled.

The theory of punctuated equilibrium developed in the 1970s and 1980s, along with debates around the notion of "species selection," was instrumental in fostering the rise of the hierarchical approach in evolutionary studies to include also the macroevolutionary levels. This theory holds that most evolutionary changes occur during short geological moments accompanied by profound modification, followed by long periods of relative stability or evolutionary stasis. Under the "microevolution = macroevolution" equation, it is argued that continual adaptive pressure (the action of natural selection) explains evolutionary change. The speciation process is, at each step, initiated, sustained, and directed by natural selection. Paleobiologists oppose this simple view by introducing the factor of "chance" or "randomness" in macroevolutionary events. Are we sure, they ask, that surviving forms are always the best adapted ones? See Raup, D. (1991), *Extinction: Bad Genes or Bad Luck?*, New York: W. W. Norton. After all, some life forms may just be lucky enough to see the extinction of some of their competitors—under a major climate shift associated with a mass extinction, for instance—allowing the survivors to thrive and diversify in any number of species, genera, and families, irrespective of the fact that they may not have been the best adapted to face the new environmental context. Differently stated, a continual or sustained adaptive pressure may not be the only, or even the predominant, factor responsible for the diversification or extinction of forms.

Under this latter view, emergent properties appear as one climbs the biological levels of the hierarchy, properties that may not necessarily be reducible to microevolutionary processes. In fact, the reverse may well be true: macroevolutionary events could impose themselves upon microevolutionary processes. Indeed, if whole groups of forms at high taxonomic levels (species, genera, and families) survive because of pure chance, their mere persistence will alter how the microevolutionary processes accompanying them will work out, irrespective of the selective value of the features involved, considering that they were not necessarily the most fit relative to both the other forms (that are now extinct because of bad luck) and the environmental context during climate shift. Evolutionary contingency, it seems, weighs heavily on evolution. The history of life, therefore, may not only be about the best adapted forms. On hierarchical thinking in evolutionary studies and other relevant texts, consult Eldredge, N. (1985), *Unfinished Synthesis: Biological Hierarchies and Modern Evolutionary Thought*, Oxford: Oxford University Press; Salthe, S. (1985), *Evolutionary Hierarchical Systems*, New York: Columbia University Press; Vrba, E. and S. J. Gould (1986), "The Hierarchical Expansion of Sorting and Selection: Sorting and Selection Cannot Be Equated," *Paleobiology*, 12: 217–28; Gould, S. J. (1989), *Wonderful Life: The Burgess Shale and the Nature of History*, New York: W. W. Norton; Ereshefsky, M., ed. (1992), *The Units of Evolution: Essays on the Nature of Species*, Cambridge: MIT Press; Gould, S. J. (2002), *The Structure of Evolutionary Theory*, Cambridge, MA: Belknap Press.

This area of modern evolutionary biology is thus much in debate, in particular as far as the role of selective factors in evolution is concerned. On the one hand, some evolutionists stress the importance of such selective factors, especially in the behavioral sciences (sociobiology). Consult Wilson, E. O. (1975), *Sociobiology: The New Synthesis*, Cambridge MA: Harvard University Press; Dawkins, R. (1976), *The Selfish Gene*, Oxford: Oxford University Press; Alcock, J. (2017), "The Behavioral Sciences and Sociobiology: A Darwinian Approach," in R. G. Delisle (ed), *The Darwinian Tradition in Context: Research Programs in Evolutionary Biology*, 37–59, Switzerland: Springer Nature. On the other hand, some evolutionists downplay the relative significance of natural selection in evolution. Consult Kauffman, S. (1993), *The Origins of Order: Self-Organization and Selection in Evolution*, Oxford: Oxford University Press; Goodwin, B. (1994), *How the Leopard Changed Its Spots: The Evolution of Complexity*, Princeton: Princeton University Press; Lewin, R. (1999), *Complexity: Life at the Edge of Chaos*, 2nd edn, Chicago: University of Chicago Press; Reid, R. (2007), *Biological Emergences: Evolution by Natural Experiment*, Cambridge MA: MIT Press.

For suggested readings concerning the patterns and processes of evolution presented in this chapter, it will be useful to suggest specific resources for each topic. (1) **Anagenesis**: See Simpson, G. G. (1953), *The Major Features of Evolution*, New York: Columbia University Press; Futuyma, D. (1987), "On the Role of Species in Anagenesis," *The American Naturalist*, 13(3): 465–573; Allmon, W. (2017), "Species, Lineages, Splitting, and Divergence: Why We Still Need Anagenesis and Cladogenesis," *Biological Journal of the Linnean Society*, 120: 474–9. (2) **Divergence and speciation**: Darwin's *Origin of Species* largely focuses on the exposition of the principle of divergence, although it is widely recognized that he failed to explain how the actual process of speciation (cladogenesis) could operate successfully. The issue has been debated since Darwin's time. Consult, for instance Mayr, E. (1963), *Animal Species and Evolution*, Cambridge: Belknap Press; White, M. (1977), *Modes of Speciation*, San Francisco: Freeman; Eldredge, N. (1985), *Time Frames: The Rethinking of Darwinian Evolution and the Theory of Punctuated Equilibria*, New York: Simon and Schuster; Paterson, H. (1985), "The Recognition Concept of Species," in E. Vrba (ed), *Species and Speciation*, 21–9, Pretoria: Transvaal Museum Monograph, No. 4; Strotz, L. and A. Allen (2013), "Assessing the Role of Cladogenesis in Macroevolution by Integrating Fossil and Molecular Evidence," *Proceedings of the National Academy of Sciences of the United States of America*, 110 (8): 2904–9; Felix, F., S. Trewick, and M. Morgan-Richards (2017), "Speciation through the Looking-Glass," *Biological Journal of the Linnean Society*, 120: 480–8. (3) **Reticulate evolution**: In the *Origin of Species*, Darwin tries to promote a theory of evolution based on the principle of divergence. Yet, unwillingly or unconsciously, he provides a lot of material supporting reticulate evolution (as will be discussed in Chapter 5 of our book). See also Lotsy, J. (1916), *Evolution by Means of Hybridization*, The Hague: Martinus Nijhoff; Grant, V. (1953), "The Role of Hybridization in the Evolution of the Leafty-Stemmed Gilias," *Evolution*, 7: 51–64; Arnold, M. and N. Fogarty (2009), "Reticulate Evolution and Marine Organisms: The Final Frontier?," *International Journal of Molecular Sciences*, 10: 3836–60; Gontier, N., ed. (2015), *Reticulate Evolution*, Switzerland: Springer Nature; Morrison, D. (2016), "Genealogies: Pedigrees and Phylogenies Are Reticulating Networks Not Just Divergent Trees," *Evolutionary Biology*, 43: 456–73. (4) **Parallelism and** *convergent evolution:* The ideas of parallel and convergent evolution have long been recognized in the field of human evolution. See Bowler, P. J. (1986), *Theories of Human Evolution: A Century of Debate, 1844-1944*, Baltimore: Johns Hopkins University Press, 131-46 and Delisle, R. G. (2017), *Debating Humankind's Place in Nature, 1860-2000: The Nature of Paleoanthropology*, New Jersey: Pearson/Prentice Hall. For the same ideas in other areas of evolutionary biology, consult Sanderson, M. and L. Hufford (1996), *Homoplasy: The Recurrence of Similarity in Evolution*, California: Academic Press; Conway Morris, S. (2003), *Life's Solution: Inevitable Humans in a Lonely Universe*, Cambridge: Cambridge University Press; Arbuckle, K., C. Bennett, and M. Speed (2014), "A Simple Measure of the Strength of Convergent Evolution," *Methods in Ecology and Evolution*, 5: 685-93; Tristan Stayton, C. (2015), "The Definition, Recognition, and Interpretation of Convergent Evolution, and Two New Measures for Quantifying and Assessing the Significance of Convergence," *Evolution*, 69: 2140-53.

Although the continually changing milieu puts pressure on life forms, it should be noted that some are more resilient than others to these pressures. Indeed, while some lineages seem quite malleable under adaptive pressure, others tend to resist change. For instance, ginkgoes, horseshoe crabs, and coelacanths have not changed much over the ages; the blueprint or body architecture upon which they are constructed may well have been so perfectly adapted over time as to no longer require major modifications. Or, alternatively, their body plans could be so tightly integrated, internally and developmentally speaking, that any serious modification might instigate a sort of organismic breakdown, which would explain their resistance to transformation. Either way, they have carried the day by being sufficiently adapted enough to the outer world, at least up until now. Nonetheless, such cases point to the importance of internal organizing principles at work within life forms. The atomistic view of organisms that sees them as a mere collection of separate traits that can independently be modified under adaptive pressure needs to be complemented by a more holistic view appealing to the internal cohesion of organisms, a holism that resists evolutionary change and acts as an organizing principle. Consult Stanley, S. M. (1979), *Macroevolution: Pattern and Process*, San Francisco: W. H. Freeman; Schopf, T. J. M. (1984), "Rates of Evolution and the Notion of Living Foosils," *Annuel Review of Earth and Planetary Sciences*, 12: 245-92; Lidgard, S. and A. C. Love (2018), "Rethinking Living Fossils," *BioScience*, 68: 760-70.

We know that life on Earth is based on a conservative DNA molecular code, a fact that strongly supports the monophyletic thesis. At the same time, we also know that the impressive diversity of life forms seen at the phenotypic level is significantly influenced by selective factors. Inevitably, this introduces some of evolutionary analogies through the rise of similar phenotypic solutions answering similar adaptive problems. This fact points in the opposite direction, toward the polyphyletic thesis. This tension is not fundamentally resolved by the recently discovered fact that a group of genes called *Pax6* (including their developmental pathways and networks) responsible for the constitution of eyes is more widespread across the taxonomical spectrum than expected. Eventually, the conservative and widespread genotypic components must face selective pressure at the phenotypic level, irrespective of the internal and holistic principles acting on the constitution of organisms. It is at the junction of the genotypic and phenotypic levels that eyes have evolved under a myriad of designs.

This is reflected in the several typologies proposed in the literature concerned with eyes. For instance, among complex types of eyes (here excluding simple types of eyes), one can distinguish between the following designs: (1) the everse eyes in Cephalopoda; the compound eyes of insects; and the inverse eyes in Vertebrata. (2) Or between: the pigment-cup eyes of many invertebrates and the camera eyes of vertebrates. (3) Or between: the ciliary type of photoreceptor cell and the rhabdomeric type. Other distinctions are made in the literature for specific evolutionary groups: parabolic superposition eyes (swimming crab: *Portunus*); pinholes (*Nautilus*); reflecting superposition eyes (decapod crustaceans); spherical lenses (many aquatic animals); refracting superposition eyes (many nocturnal insects, crustaceans); multiple lenses (copepod crustaceans: *Pontella*, *Copilia*); afocal apposition (butterflies); corneal refraction (spiders, penguins, seals, diving birds); neural superposition eyes (dipteran flies); concave reflectors (rotifers, platyhelminths, scallops: *Pecten*, copepod crustaceans, ostracod crustacean: *Gigantocypris*); apposition eyes (found in all three arthropod subphyla).

In this mixed taxonomic context concerning the issue of eyes, it is easy to see why the monophyletism-polyphyletism question is not easily disentangled. Consult, for instance, Salvini-Plawen, L. and E. Mayr (1977), "On the Evolution of Photoreceptors and Eyes," *Evolutionary Biology*, 10: 207–63; Land, M. and R. Fernald (1992), "The Evolution of Eyes," *Annual Review of Neurosciences*, 15: 1–29; Arendt, D. (2003), "Evolution of Eyes and Photoreceptor Cell Types," *International Journal of Developmental Biology*, 47: 563–71; Oakley, T. (2003), "On Homology of Arthropod Compound Eyes," *Integrative and Comparative Biology*, 43: 522–30; Mayr, E. (2001), *What Evolution Is*, New York: Basic Books, 205–7; Gehring, W. (2004), "Historical Perspective on the Development and Evolution of Eyes and Photoreceptors," *International Journal of Developmental*

Biology, 48: 707–17; Salvini-Plawen, L. (2008), "Photoreception and the Polyphyletic Evolution of Photoreceptors (with Special Reference to Mollusca)," *American Malacological Bulletin*, 26: 83–100; Land, M. and D. Nilsson (2012), *Animal Eyes*, 2nd edn, Oxford: Oxford University Press; and Glaeser, G. and H. Paulus (2015), *The Evolution of the Eye*, Dordrecht: Springer.

Earth is an unstable place whose physical changes are often reflected in biological changes. Fields such as biogeography, geology, paleontology, and paleoenvironment, document these facts. The list of physical changes is long: the break-up of continents; their derivation on Earth's surface; the continual redesign of oceans under continental motion; the rise of islands and mountains followed by their erosion; the ever-changing course of rivers; climate changes; the advance and retreat of glaciers; fluctuations in sea levels, the transformation of barren lands into fertile areas or the reverse; the impact of meteorites and asteroids on Earth's global climate. Such abiotic changes could only have led to biotic ones by modulating the pace of biological evolution; mixing up the cohabitation of various species from time to time; isolating some other species for a time and putting them into contact again; creating the conditions for speciation, reticulate evolution, parallel evolution, convergent evolution, local extinctions, and mass extinctions. Consult Huston, M. (1994), *Biological Diversity: The Coexistence of Species on Changing Landscapes*, Cambridge: Cambridge University Press; Avise, J. C. (2000), *Phylogeography*, Cambridge MA: Harvard University Press; McCarty, D. (2009), *Here Be Dragons: How the Study of Animal and Plant Distributions Revolutionized Our Views of Life and Earth*, Oxford: Oxford University Press; De Queiroz, A. (2014), *The Monkey's Voyage: How Improbable Journeys Shaped the History of Life*, New York: Basic Books.

In closing our discussion in this chapter's bibliography, we would mention that the thinking regarding the relationship between organisms and their milieu has changed over the last few decades. For a long time (say, roughly between the 1930s and the 1980s), it was widely believed that the role of organisms in their engagement with the environment was a "passive" one: organisms could only confront the environment with the genetic variations they had inherited. On this view, the flow of biological information was seen as a one-way street, with the genes coding for the variations (information) that are expressed at the phenotypic level. It is now recognized that organisms are more "open" entities than had previously been understood, with complex feedback loops between the environment, the phenotypic level (structures and behaviors), and processes occurring at the genotypic level. It seems that, after all, the environment might be able to impact the internal processes of organisms, thus allowing biological information to flow in two directions: (1) from the genes to the confrontation with the environment (through the phenotypic level) and (2) from the environment to the genotypic level (again, through the phenotypic level). It is not yet entirely clear how this new form of "Lamarckism" will be harmoniously integrated into our modern understanding of evolutionary biology. For further discussion, consult Jablonka, E. and M. Lamb (1995), *Epigenetic Inheritance and Evolution: The Lamarckian Dimension*, Oxford: Oxford University Press; Jablonka, E. and M. Lamb (2005), *Evolution in Four Dimensions: Genetic, Epigenetic, Behavioral, and Symbolic Variation in the History of Life*, Cambridge MA: MIT Press; Gissis, S. and E. Jablonka, eds (2011), *Transformations of Lamarckism*, Cambridge MA: MIT Press; Depew, D. J. (2017), "Darwinism in the Twentieth Century: Productive Encounters with Saltation, Acquired Characteristics, and Development," in R. G. Delisle (ed), *The Darwinian Tradition in Context: Research Programs in Evolutionary Biology*, 61–88, Switzerland: Springer Nature.

2

What Time Selected from Darwin: The Standard View

The preceding chapter was intended to show that an essential feature of evolution is its openness, something readily apparent in the wide range of biological processes and patterns just considered: divergence, convergence, reticulate evolution, near-stagnation, rapid modification, and extinction. Underlying this openness is the critical need for organisms and the myriad of lineages to which they belong to adapt to the surrounding world. This unstable world itself defies prediction, and yet another layer of randomness is added when these factors are set in motion together: one cannot possibly predict how things will turn out when a set of competing lineages with random biological variations—shuffled and reshuffled with each successive generation—is reunited in a new geographical place and geological time. Evolution is fundamentally unpredictable. This indeterminacy stands sharply at odds with the equally deep human desire for prediction. While evolutionary theories proposed since the nineteenth century have not, generally speaking, acknowledged the openness and unpredictability of evolution (see the Annotated Bibliography at the conclusion of this chapter for greater detail), those that have dominated the scientific scene since the 1950s have done so to a significant degree. Because it is often assumed Charles Darwin embraced the openness of evolution, it is this assumption regarding his theory that constitutes the primary investigative tool of our book.

Appreciating the breathtaking complexity of evolution in full has taken researchers some time, requiring, as it has, a dramatic change in perspective. This realization had also to await the steady accumulation of empirical data and fossil evidence we now possess, much of which was simply unavailable in Darwin's day. Nonetheless, many scholars today believe that Darwin did manage to capture the larger part of this complexity in the various editions of the *Origin of Species* (1859–1872), arguing further that his theory of evolution was his deliberate attempt to make sense of it. Many statements in the *Origin* seem to back this up: for instance, Darwin repeatedly (and famously) reminds his readers that there is "no fixed law of development," a maxim he presents in several different guises, implicitly and explicitly. This constitutes for many scholars a declaration of his faith in the notion we have ourselves have taken pains to emphasize in the foregoing chapter as essential: that evolution is a truly *open* process. In making this claim, they argue, Darwin was very much going against the grain of his time, sharply opposing the idea of evolution as internally organized, predetermined, applicable in equal measure to all lineages, and propelled by its own internal momentum independently of environmental pressure. Darwin writes:

> These several facts accord well with my theory. I believe in *no fixed law of development*, causing all the inhabitants of a country to change abruptly, or simultaneously, or to an equal degree. The process of modification must be extremely slow. The variability of each species is quite independent of that of all others. Whether such variability be taken advantage of by natural selection, and whether the variations be accumulated to a greater or lesser amount, thus causing a greater or lesser amount of modification in the varying species, depends on many *complex contingencies*,—on the variability being of a beneficial nature, on the power of intercrossing, on the rate of breeding, on the slowly changing physical conditions of the country, and more especially on the nature of the other inhabitants with which the varying species comes into competition. Hence it is by no means surprising that one species should retain the same identical form much longer than others; or, if changing, that it should change less. (Darwin, 1859, p. 314) [our emphasis]

Understandably, scholars of today find it hard not to read statements such as these as proof positive of Darwin's modernity. This is reflected in the following passages from two major biologists of the second half of the twentieth century who, unquestionably, have influenced the interpretation of historians. In *The Growth of Biological Thought* (1982), Ernst Mayr writes:

> Interestingly, Darwin's approach was completely in line with modern theory. He realized he could never demonstrate evolutionary conclusions with the certainty of a mathematical proof. Instead, in about twenty different parts of the *Origin* he asks: "Is this particular finding—whether a pattern of distribution or an anatomical structure—more easily explained by special creation or by *evolutionary opportunism*?" Invariably, he insists that the second alternative is the more probable one. Darwin anticipated many of the most important tenets of the current philosophy of science.[1] [our emphasis]

Mayr then goes on to explain that, with the publication of the *Origin of Species*, a new conception of science had to be developed, one adapted to confront unique and unpredictable evolutionary events in the argumentative style of historical narratives. In different words, yet in a similar spirit, Stephen Jay Gould writes in *Wonderful Life* (1989):

> Replay the tape [of life] a million times from the Burgess beginning [in the Cambrian period more or less 500 million years ago], and I doubt that anything like *Homo sapiens* would ever evolve again ... When we set our focus upon the level of detail that regulates most common questions about the history of life, contingency dominates and the predictability of general form recedes to an irrelevant background. Charles Darwin recognized this central distinction between *laws in the background* and *contingency in the details*.[2] [italics original]

Reasonable though such interpretations may be, it is worth noting that Darwin's words are often misleading, or rather, imprecise enough so as to permit modern readers to read much of their own hard-won knowledge and theoretical assumptions back into this nineteenth-century text. Indeed, schooled in twentieth- and twenty-first-century biology, the modern reader finds it all too easy to project current understandings onto the nineteenth century, including many of the threads discussed in Chapter 1. At this point, we will lay our cards on the table: a hard look at the *Origin* reveals that, surprising as it may sound, Darwin was *not* a firm believer in *evolutionary contingency or opportunism* as defined today and in fact stopped well short of taking this modern precept on board. Applying this critical reading will allow us in later chapters to highlight the many other neglected elements in the *Origin of Species* that cast doubt on the standard interpretation and help us understand why the real, somewhat old-fashioned, Darwin has largely eluded us.

In charging modern scholars with anachronism, we are keenly aware that they have been aided by Darwin himself, sometimes unintentionally (but not always), by the way he presents his ideas in his masterwork. In the *Origin*, as already alluded to in our "Introduction" and as will be more clearly seen in the following chapters, Darwin employs an argumentative style that tends to paper over a number of doubts, tacit assumptions, inconsistencies, and contradictions. This fact has been openly recognized by both friends and foes of Darwin. For instance, George Levine writes, in 1988: "[Darwin] had the power to imagine what wasn't there and what could never be seen, and he used analogies and metaphors with subtlety and profusion as his imagination actually defied the experience that Baconian theory privileged."[3] For his part, Ernst Mayr, a self-proclaimed Darwinian if there ever was one, writes, in 1985: "I want to issue a warning at the outset. Darwin was a great pioneer, a person with an exceptionally fertile mind, but like other fertile thinkers, he had considerable trouble sticking to a consistent 'party line'. On almost any subject he dealt with— and this includes almost all of his own theories—he not infrequently reversed himself."[4] Similarly, Jacques Barzun writes in 1958:

> A worse fault than obscurity, in view of the wide faith accorded by the nineteenth century and ours to scientific works, is Darwin's hedging and self-contradiction; for it enabled any unscrupulous reader to choose his text from the *Origin of Species* ... with almost the same ease of accommodation to his [her] purpose as if he [she] had chosen from the Bible.[5]

And in a substantially less sympathetic tone, Samuel Butler complains in 1911:

> I assure the reader that I find the task of forming a clear, well-defined conception of Mr. Darwin's meaning, as expressed in his "Origin of Species," comparable only to that of one who has to act on the advice of a lawyer who has obscured the main issue as far as he [she] can, and whose chief aim has been to make as many loopholes as possible for himself [herself] to escape through in case of his [her] being called to account.[6]

Lest Darwin's foes be given the last word on this, the reader is also invited to read the 1860 remarks (quoted in full in our "Conclusion") of Darwin's contemporary and friend, Thomas Henry Huxley. He, too, is clearly struck by Darwin's convoluted style of argumentation.

Despite its apparent simplicity and clarity, the *Origin of Species* must be approached with deliberate caution. Indeed, scientific-minded readers more concerned with uncovering the truth of evolution per se than with divining Darwin's line of argument may be more likely to be deceived by the *Origin*. This phenomenon is readily apparent throughout the multitude of versions of evolutionary theory appearing more or less continuously since the publication of the *Origin*, such as those, for instance, known variously as Neo-Darwinism, the Modern Synthesis, or the Extended Evolutionary Synthesis.

At this point, full disclosure is in order, and a *mea culpa*: lest the reader think our critical approach comes from a position of presumed superiority, we hasten to add that we are in no way sparing ourselves. As alluded to earlier, we for many years entirely overlooked these contradictions and alternative understandings, not only on our first readings of Darwin, but for years afterward. Once these contradictions became apparent, however, they prompted a kind of gestalt shift and became impossible to ignore. Reading the *Origin* now, what strikes us most is the degree to which our (perhaps natural) tendency to seek founding figures has distorted our collective understanding of Darwin.

The Post-Darwinian Theory of Evolution

Since the 1930s, a number of evolutionists have worked to erect a theory of evolution built around a mechanism that selects the "best" variations, discarding the others. The details of this theory are familiar enough: the rise of genetic variations at the phenotypic (i.e., observable) level provides the raw material, with the sorting of these variations occurring through the readily familiar process of *natural selection*, a process that follows from the competition that naturally ensues when life forms confront each other and their environment. Only those organisms fortunate enough to have the "right" variations survive and reproduce. The quest to develop a comprehensive theory of evolution based on natural selection led many to turn to Darwin himself. Drawing inspiration, perhaps, from the full title of his 1859 book, *On the Origin of Species by Means of Natural Selection, or the Preservation of Favoured Races in the Struggle for Life*, twentieth-century evolutionists saw themselves as Darwin's rightful heirs, extending his work by following directly in his footsteps, unaware they were introducing novel elements into the original theory and departing from it significantly at various points along the way (see our Chapter 7 for a summary). In many respects, then, the so-called neo-Darwinian theory they developed might more accurately be termed "post-Darwinian," acknowledging the important differences between Darwin and his successors. This will be our choice of term for the remainder of the book.

The post-Darwinian theory of evolution places natural selection at its very core. As the central explanatory mechanism of this theory, natural selection is responsible for generating *evolutionary contingency*; that is, its role is to ensure evolution remains precisely as sketched above: an unpredictable, open-ended affair. Biological evolution is envisioned as the interaction of two independent and fundamentally random chains of events: (1) the unpredictable rise of variations among biological populations; (2) the unpredictable changes in the organic and inorganic milieu. Standing at the crossroads of these two strands, natural selection acts as a kind of mediator, a sorting mechanism standing between those organisms and biological populations that find themselves with the right features, in the right place, at the right time, and so have the chance to stay in the game, and those that do not, and so fall by the wayside into extinction. The overall picture of evolution that emerges from this account is that of a sum of independent and unpredictable pathways traced out by innumerable lineages. Under this view, evolution leaves no room for predetermination or any other overarching process applicable to all lineages or capable of directing their course irrespective of external conditions. Each lineage meanders along its own unique path in response to its own unique circumstances.

In recent decades, post-Darwinians have come to read the *Origin of Species* as presenting a coherent and, and above all, comprehensive theory organized around the core concept of natural selection, with sufficient explanatory power to unify a vast array of evolutionary phenomena known to us. Differently stated, they argue that Darwin offered to posterity a well-structured theory organized around a causal core (natural selection), that is, an evolutionary mechanism explaining a whole range of phenomena. Under their construal of the book, the first chapters of the *Origin* are devoted to supporting the empirical foundations of natural selection as a force capable of initiating biological change, with later chapters putting this mechanism into practice, functioning only to breathe meaning into discoveries in fields as diverse as biogeography, paleontology, taxonomy, morphology, and embryology.

While it certainly seems right to acknowledge that Darwin expended some effort to present his theory of evolution along the lines just outlined, the question is—and this is where we part company with the aforementioned scholars—whether he really succeeded. Here it is important to distinguish between what Darwin set out to achieve, on the one hand, and how far he actually progressed toward this goal, on the other. To make this assessment we can't simply take Darwin's word for it: instead, we must revisit the *Origin of Species* with a critical eye. It is our contention that post-Darwinian efforts to cast Darwin in a modern light do not survive this additional scrutiny, for, far from offering a well-structured, coherent, or unifying theory, the pages of the *Origin* provide claims, which, on our view, actually present a rather different view of biological evolution. To run ahead of our story here, we hold that the numerous tensions and contradictions already registered by Darwin scholars (such as those reviewed above) testify to an unstable explanatory system erected at the junction of two competing, and mutually inconsistent, worldviews: the static and the evolutionary. We hope to wring a different kind of order out of this apparent tension in the *Origin*, but before doing so we need to present the *Origin* as it is construed by the post-Darwinians, if only to follow through on the suggestions we have only hinted at so far.

At the outset, we would acknowledge the seductive appeal of seeing Darwin's magnum opus as characterized by its internal coherence and power of explanatory unification. Indeed, evolutionary biologists had themselves long desired to see their field as an area of science grounded in a solidly unified theory. Ever since 1687, the year Isaac Newton managed to tie together a range of seemingly unrelated natural phenomena—falling objects, cycles of tides, trajectories of planets—by means of the unifying power of the theory of universal gravitation, his scientific heirs in other fields have, understandably, aimed to follow in his footsteps. Despite these ambitions, however, replicating Newton's breathtaking achievement in physics has thus far proven elusive in the complex field of biology, for Darwin and his successors both.

What Was Selected from the *Origin of Species*

Before proposing an alternative (and what we believe is a more accurate) understanding of Charles Darwin, however, we should first identify the specific elements comprising the *standard view* and its accompanying analytical framework. Though the details of the view may be familiar to many readers, laying them bare at the outset will help establish how and where the post-Darwinians go wrong.

The post-Darwinian account takes the first five chapters of the *Origin* to be laying out a theory of biological evolution with the mechanism of natural selection at its explanatory core. In support of this thesis, post-Darwinians claim to find the following components in the *Origin*:

1. Although little is understood of the origin, expression, and redistribution of biological variations through reproduction, organisms and populations are seen as the carriers of these variations.
2. The action of natural selection in the wild is conceived on analogy with the kind of work done by plant and animal breeders. Just as breeders tap into the reservoir of variations of populations of domestic animals and plants by selecting some of them in order to modify and create new breeds, natural selection sorts out the organisms and populations that are best fitted to carry on. Over time, the resulting changes produce adaptations among surviving forms and lead the rest to extinction.
3. The selection process just described is inexorable, the result of two diametrically opposed factors: whereas the resources required for survival (especially food, climate, and physical space) are always limited, the natural tendency of living things is always to reproduce. More organisms are generated than can possibly survive, so only the carriers of the best

variations will live to reproduce. Here, one might say that evolution happens when the irresistible force of reproduction meets the immovable object of environment.

4. More specifically, the struggle for existence (survival and reproduction) occurs in the context of limited "places" in nature. No organism or population can carve out a niche *everywhere* given the vast range and diversity of potential habitats to be found on planet Earth: under water, below ground, in the air; climates ranging from extreme, unremitting cold at the poles and high altitudes to sweltering heat in the equatorial and tropical zones. Similarly, no organism or population can exploit all potential resources the Earth provides; some food sources, for example, are simply off the table for a given organism. Life forms are adapted to occupy *particular* natural settings and circumstances. "Niches" is indeed an apt term, since inhabiting these spaces successfully demands much more than occupying a given physical space, including the organism's inborn capacities to exploit its surroundings. From the perspective of organisms and populations, nature is divided into those areas with a few niches open to them and the many more entirely closed off to them. The sum and distribution of niches considered as a whole is known as the *economy of nature*.

5. In a world characterized by the limited niches making up the economy of nature, any variations that enable organisms and populations to beat out the competition will increase the survival chances of their possessor (and the traits themselves). These competition-enhancing features constitute a dynamic force that drives evolutionary change, a drive that mainly expresses itself by *avoiding* competition: diverging or evolving away from competitors reduces the level of competition. We would thus expect to see a pattern of divergence predominate throughout as time goes on.

The overall picture of evolution, under this thesis, is one organized around common ancestors whose direct descendants continue to fan out over time (divergence). In this process, an original "parent" population eventually splits into two distinct "daughter" populations through *cladogenesis*—each of which then evolves away from both their common parent population and each other— (its "sisters")—becoming ever more distinct from one another. As this continues over geological time, at some point the daughter populations are different enough in kind to deserve their own category (or new name). Since at the outset the parent population formed a more or less unitary whole, the organisms belonging to it are given the same name. When that single population begins to split, subgroups arise within it. At first, the two subgroups bear relatively few differences, at which point they are best thought of as two distinct *varieties* within the same original population. As the splitting process continues over time, differences accumulate. What were once two different *varieties* now become two different *species*. If the differentiation process persists, the two *species* will eventually need to be reclassified as separate *genera*. Further differentiation will divide them into *families*, perhaps even *orders*.

Forms that develop along the lines just described are said to be *climbing the taxonomic scale* (*taxonomy* being the study of classification of life forms; see Figure 8). It has long been recognized that life forms are not randomly organized but come in obvious clusters: donkeys, horses, and zebra share readily observable similarities, as do cat-like animals (pumas, lynx, and tigers) and dog-like animals (wolves, coyotes, jackals). Of course, these similarities are explained by shared ancestry: horses, cats, and dogs all evolved from separate common ancestors. The more these forms evolved away from their respective common ancestors, the more different from one another they became. This means different taxonomic categories must be created for each. Closely related forms are assigned to the lower taxonomic levels of varieties and species; more remote but still related

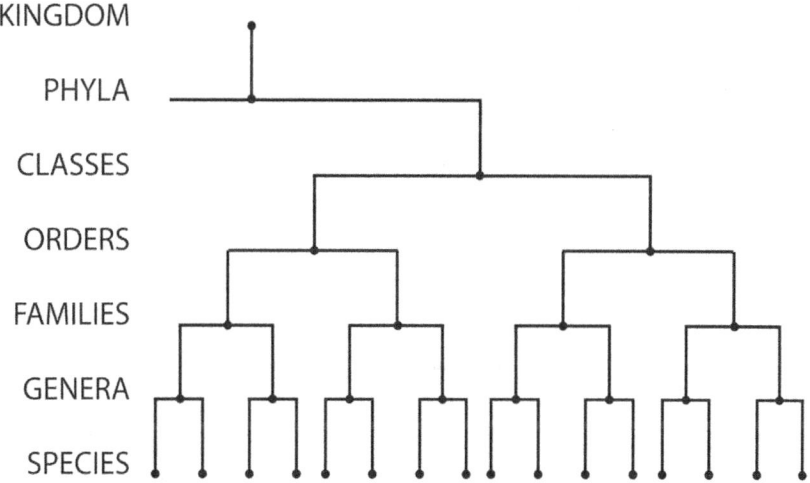

Figure 8 Taxonomic scale. Life forms are organized under a system founded on the principle of entities classified into ever more inclusive categories, a system ultimately expressing something about the genealogical relationships of such entities. The main levels of the system are the following: species, genera, families, orders, classes, phyla, and kingdom.

forms are classified into separate genera. As forms become ever more distinct, genealogically speaking, they move up the taxonomic scale, to the family level, the order level, and even the class level (for very distant forms).

To recapitulate: the components just presented comprise the *standard view*, the theoretical elements typically attributed to Darwin in the first five chapters of the *Origin of Species*. Contemporary scholars usually interpret Chapters 6 to 8 as Darwin's attempt to handle the problem of the existence of intermediate forms. Indeed, the would-be evolutionist now faces a challenge to her theory, however. If species are constantly changing along the lines described above, and this force is immanent and ubiquitous as conceived, we should be able to see evidence of these forces at work in nature: traces of these transformations or other evidence of intermediate or transitional evolutionary links connecting forms. At a minimum, abundant evidence would tend to confirm this hypothesis (a lack of evidence, while not in itself fatal, would not inspire confidence in the theory). Darwin clearly understood this and set about surveying currently living forms in search of behaviors and anatomical structures showing promise of being evolutionary bridges.

Closely related groups belonging to different *varieties*, *species*, and *genera* are not thought to be sharply separated from one another from the standpoint of reproduction. Instead, they are seen as falling along a continuum that ranges from nearly perfect fertility (as seen between close varieties) to total sterility (as seen between affiliated families). This conception is consistent with the hypothesis that when closely related lineages diverge from a common ancestor, they simultaneously climb the taxonomic scale. Natural selection is thus envisioned as a force powerful enough to build complex features out of a simple initial state, merely by tinkering with existing biological variations.

In Chapters 9 and 10 of the *Origin*, Darwin takes up paleontology and the fossil record, a move proponents of the *standard view* regard as undertaken wholly in the service of advancing a theory of evolution based on divergence. One way to verify the divergence hypothesis is by backtracking along the pathways taken by evolution, digging back into the geological past. As we get closer and closer to the presumed common ancestor, we should see the now-distinct but related forms coming to resemble each other to an increasing degree. Following on this course, we should see forms converging on the original, undifferentiated parent form. This is precisely what Darwin claims to have discovered in his own paleontological efforts, though he seems also to have realized that the fossil record was not as informative and as supportive as he had perhaps hoped.

Confronted with this lack of strong evidence and a potentially fatal threat to his overall account, Darwin goes to great pains in these chapters to provide what are, in effect, attempts at explaining away this counterfactual evidence or apparent "shortcomings" of the fossil record on this score (some of which will be presented in our Chapter 3).

Judged from the perspective of the *standard view*, Chapters 11 and 12 of the *Origin of Species* serve to postulate a single birthplace, or "cradle," from which closely related forms emerge. Drawing on the field of biogeography—the study of the distribution of life forms across geographical space—Darwin tries to explain how a number of currently living forms have managed to spread around the globe to the locales they now inhabit, an endeavor that is consistent with his overall tendency to emphasize "divergence." If affiliated forms arose out of a common ancestor through divergence before migrating elsewhere, he claims, it would seem feasible to look for their original home. Darwin pursues this strategy with the assistance of two basic suppositions. First, we may be able to glean information about the location of each "cradle"—and the possible path back to the common ancestor—by keeping in mind that closely related *varieties* and *species* tend to replace each other as one observer travels across the geographical space they inhabit. For instance, imagine three varieties of a plant living in a given area, say, an island. The first of these varieties (let's call it *Type A*) inhabits the southern part, the second (*Type B*) the middle part, and the third (*Type C*) the northern part of this same island. In this example, Type B occupies an intermediate geographical location, but it would also seem to hold a middling position genealogically, in virtue of the intermediate features it shares with Types A and C. Darwin also supposes that *varieties* and *species* that come to be arranged in this way may also come to be modified somewhat by the prevailing geography, especially by obstacles such as lakes, mountains, and other features of the landscape, which may act to arrest and channel their movements and, simultaneously, alter the evolutionary pressures ultimately instigating various adaptations.

Lastly, Chapter 13 of the *Origin*, when looked at through the lens of the *standard view*, is taken by post-Darwinians as a synthesis of work in several distinct fields, each in the service of supporting Darwin's core theoretical commitments laid out in Chapters 1–5. In these final pages, Darwin explores the disparate fields of taxonomy, morphology, embryology, and comparative anatomy (the study of rudimentary organs). An exploration of the field of taxonomy forms the heart of the chapter, from which Darwin draws on the notion that clusters of life forms (the cat- or horse-like animals mentioned above) should be organized and classified in accordance with groups of affiliated forms bound in common ancestry. The other fields mentioned above are then brought in largely to support this central idea. Morphology, for instance, seems to suggest that beyond their obvious differences, vertebrates such as mammals, birds, reptiles, and fish—now separated at the level of class—are nonetheless united by a common basic architectural structure: a vertebral column. For its part, embryology provides us with a way of glimpsing the vertebrates' common ancestor based on the supposition that in their early embryological stages they retain something of the common ancestral condition, a condition that has long since been erased by adaptive pressure in adult forms. Comparative anatomy helps us see that features no longer useful for the survival of forms sometimes persist in an atrophied or vestigial aspect. These are useful data for determining earlier relationships between populations, as they represent dead ends, traces, and leftovers, all evidence of earlier genealogical connections.

Conclusion

Presented in this way, the standard view of the *Origin of Species* does look rather impressive in terms of its coherence and comprehensiveness. One gathers that Charles Darwin has indeed presented a theory of evolution capable of making sense of a wide range of evolution's complexities and explanatory challenges, a theory believed to contain the following notions: (1) evolutionary change and adaptation resulting from unpredictable biological variations encountering unforeseeable challenges; (2) a differentiation process (divergence) impacting lineages, but to a greater or lesser extent according to various adaptive contexts; (3) the changing pace of evolutionary change in lineages following the modulation of adaptive pressure from time to time and place to place. These proposals seem to fall neatly together under the oft-repeated motto, "no fixed law of development": that is, evolution cannot be a predetermined process equally applicable to all lineages independently of external conditions; rather, it is a process that involves as many independent evolutionary outcomes as there are different lineages, since each faces its own unpredictable challenges equipped with its own unique set of strengths and weaknesses.

Unfortunately, this cozy picture ultimately reveals itself to be rather incomplete, as we hope to show in coming chapters. Darwin, it seems, was not nearly as modern as the now-standard construal of his thought would suggest. In particular, we will show that the explanatory components listed above—which, we acknowledge, are indeed there to be found in the *Origin*—take on quite different meanings when interpreted in light of other features often left out of the official story favored by post-Darwinians.

Annotated Bibliography

As explained in the Annotated Bibliography for our Chapter 1, some theories focus more on a holistic approach to life (as opposed to an atomistic one), a feature assumed to be associated with forms resisting change. The evolutionary process might, therefore, be less open than believed on the assumption that the cohesiveness of organisms and species might contribute to channel evolution more than some theories care to recognize. For a historical overview of theories stressing the oriented nature of evolution, consult Bowler, P. J. (1983), *The Eclipse of Darwinism: Anti-Darwinian Evolution Theories in the Decades around 1900*, Baltimore: Johns Hopkins University Press. For more recent accounts based either on interpreting the fossil record or on immanent self-organizing principles, consult Conway Morris, S. (2003), *Life's Solution: Inevitable Humans in a Lonely Universe*, Cambridge: Cambridge University Press and Kauffman, S. (1995), *At Home in the Universe: The Search for the Laws of Self-Organization and Complexity*, Oxford: Oxford University Press. Again, because the *standard view* typically credits Darwin's theory with an open and unpredictable view of evolution, this book will focus only on those theories appealing to "evolutionary contingency."

Charles Darwin produced six editions of the *Origin of Species* between 1859 and 1872. The claims about Darwin made here pertain to all editions. For the sake of simplicity, however, with very few exceptions, our case will be based on the first edition of 1859. An interested reader will find a systematic exposition of the differences that exist between the various editions in Darwin, C. R. (1959), *The Origin of Species: A Variorum Text*, M. Peckham (ed), Philadelphia: University of Pennsylvania Press. For an analysis of some of these differences, see Vorzimmer, P. (1970), *Charles Darwin: The Years of Controversy*, Philadelphia: Temple University Press; Liepman, H. (1981), "The Six Editions of the Origin of Species: A Comparative Study," *Acta Biotheoretica* 30 (3): 199–214; and Hoquet, T. (2013), "The Evolution of the *Origin*," in M. Ruse (ed), *The Cambridge Encyclopedia of Darwin and Evolutionary Thought*, 158–64, Cambridge: Cambridge University Press.

For those pages apparently supporting Charles Darwin's notion of "no fixed law of development," see especially Darwin, R. (1859), *On the Origin of Species*, London: John Murray, 313–15, 318, 331–2, 343, 351, 408–9.

For the classical exposition of Darwin's presumed commitment to evolutionary contingency as well as the commitment to a fully open-ended account of evolution adopted by contemporary evolutionists, consult Simpson, G. G. (1964), *This View of Life*, New York: Harcourt, Brace & World, 176–89; Mayr, E. (1982), *The Growth of Biological Thought*, Cambridge: Belknap Press, 21–82; Gould, S. J. (1986), "Evolution and the Triumph of Homology, or Why History Matters," *American Scientist*, 74: 60–9; Gould, S. J. (1989), *Wonderful Life: The Burgess Shale and the Nature of History*, New York: W. W. Norton, 277–91; Gould, S. J. (2002), *The Structure of Evolutionary Theory*, Cambridge MA: Belknap Press, 1332–6.

The thesis that Darwin presented a well-structured theory organized around a causal core (natural selection) giving meaning to a host of disciplines has been proposed in Ruse, M. (1975), "Darwin's Debt to Philosophy: An Examination of the Influence of the Philosophical Ideas of John F. W. Herschel and William Whewell on the Development of Charles Darwin's Theory of Evolution," *Studies in History and Philosophy of Science*, 6: 159–81; Ruse, M. (1979), *The Darwinian Revolution*, Chicago: University of Chicago Press; Ruse, M. (2000), "Darwin and the Philosophers: Epistemological Factors in the Development and Reception of the Theory of the *Origin of Species*," in R. Creath and J. Maienschein (eds), *Biology and Epistemology*, 3–26, Cambridge: Cambridge University Press; Hodge, J. (1977), "The Structure and Strategy of Darwin's Long Argument," *British Journal for the History of Science*, 10: 237–46; Hodge, J. (1989), "Darwin's Theory and Darwin's Arguments," in M. Ruse (ed), *What the Philosophy of Biology Is*, 163–82, Dordrecht: Kluwer.

An overview of what modern scholars usually extract from the *Origin of Species* under the *standard view* is laid out in a clear and compact fashion in Ruse, M. (1999), *The Darwinian Revolution*, 2nd edn, Chicago: University of Chicago Press, 188–98; Bowler, P. J. (1990), *Charles Darwin: The Man and His Influence*, Oxford: Basil Blackwell, 114–25.

Against the received view, we argue here that *direct* intellectual connections between the *Origin of Species* and twentieth- to twenty-first-century developments are far weaker and less obvious than usually believed. For details of this thesis, consult Delisle, R. G. (2017), "From Charles Darwin to the Evolutionary Synthesis: Weak and Diffused Connections Only," in R. G. Delisle (ed), *The Darwinian Tradition in Context: Research Programs in Evolutionary Biology*, 133–67, Switzerland: Springer Nature. Among a long list of authors promoting the thesis that such strong connections exist, see Provine, W. (1985), "Adaptation and Mechanisms of Evolution after Darwin: A Study in Persistent

Controversies," in D. Kohn (ed), *The Darwinian Heritage*, 825–66, New Jersey: Princeton University Press; Depew, D. and B. Weber (1995), *Darwinism Evolving*, Cambridge: MIT Press; Gayon, J. (1998), *Darwinism's Struggle for Survival: Heredity and the Hypothesis of Natural Selection*, Cambridge: Cambridge University Press; Ruse, M. (2011), "Is Darwinism Past Its Sell-By Date? The *Origin of Species* at 150," *Studies in History and Philosophy of Biological and Biomedical Sciences*, 42: 5–11.

Notes

1 Mayr, E. (1982), *The Growth of Biological Thought*, Cambridge: Belknap Press, 27.
2 Gould, S. J. (1989), *Wonderful Life: The Burgess Shale and the Nature of History*, New York: W. W. Norton, 289–90.
3 Levine, G. (1988), *Darwin and the Novelists*, Chicago: University of Chicago Press, 1.
4 Mayr, E. (1985), "Darwin's Five Theories of Evolution," in D. Kohn (ed), *The Darwinian Heritage*, Princeton: Princeton University Press, 756.
5 Barzun, J. (1958), *Darwin, Marx, Wagner: Critique of a Heritage*, 2nd edn, New York: Doubleday, 75.
6 Butler, S. (1911), *Evolution, Old and New: Or the Theories of Buffon, Dr. Erasmus Darwin and Lamarck, as Compared with That of Charles Darwin*, 2nd edn, London, 381.

Part Two

Charles Darwin and the Static Worldview

3

The Tree That Hides the Forest: Charles Darwin's "Tree of Life"

For many commentators on Darwin, the centerpiece of the *Origin of Species* is its sole illustration (sometimes referred to simply as the *"Diagram"*), which is thought to encapsulate his view of evolution in visual form (see Figure 9). Darwin's Diagram shows the evolution of eleven distinct lineages (from A to L) over the course of geological time. A cursory look suggests an overall picture that is in the main consistent with the modern view of evolution: its branches suggest the full expression of a free and open evolutionary process governed by *evolutionary contingency* (as explained in our previous chapter). Indeed, the Diagram seems to capture the key manifestations one would expect of this kind of evolution:

1. It appears life has diversified in a haphazard fashion, with many evolutionary lines diverging from one another to various degrees, as should be expected under changing and unpredictable biotic and abiotic conditions.
2. Life forms are organized around common ancestors.
3. While many lines diverge, some others move toward each other (as seen when some members of lineages A and I are compared, for instance).
4. Many lines have gone extinct because they could not find evolutionary solutions to threatening challenges, just as one would predict given life's perpetual encounters with unpredictable conditions.
5. On the other hand, some lineages were apparently so well adapted to conditions that they remained perfectly stable for various lengths of time, until most eventually became extinct, as seen in lineages B to H and K and L.
6. Finally, when one projects the common ancestors of lineages A to L further back in geological time, it can be seen that they tend to get closer and closer to each other, as if all had sprung from a common ancestor of some sort.

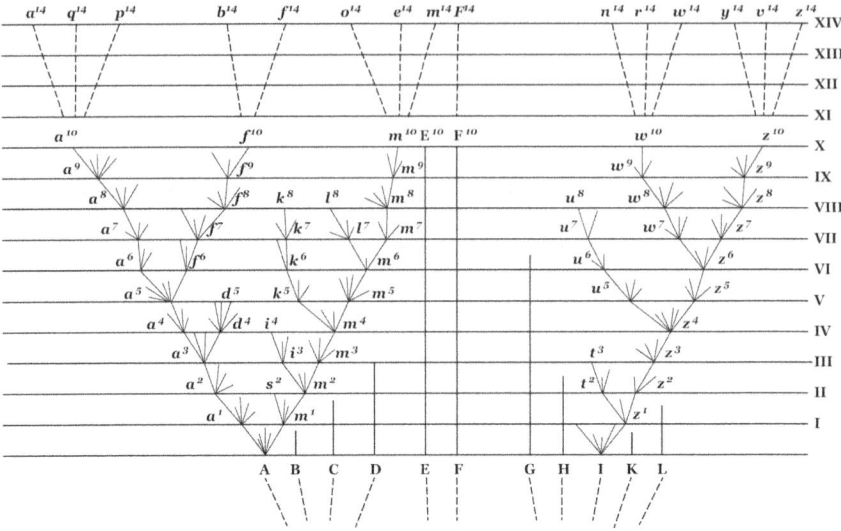

Figure 9 Darwin's Diagram. This is the sole illustration to be found in the *Origin of Species*. This reveals that Darwin believed the principle of divergence constitutes the main evolutionary process-pattern accompanying evolutionary change.

Modern scholars have taken these features of the Diagram to be compelling evidence and thus a reason to cast Darwin in the role of first truly modern evolutionist, and it is easy to see why. Nonetheless, Darwin's Diagram is one place among many in the *Origin of Species* where it is all too easy to project our modern perspectives. It will be our task in this and following chapters to challenge the standard view by bringing to light the abundant evidence to the contrary, offered by Darwin himself. Things have been said in the name of Darwin that cannot be attributed to him. Indeed, Darwin's so-called tree of life hides a thicket of quite unmodern ideas, many of them inherited from the static worldview prevailing in the seventeenth and eighteenth centuries. The main challenge issuing from Darwin's Diagram, we hold, comes from approaching it too directly, as it takes on a different meaning when seen against the textual background provided in the *Origin of Species*. With these precautions in mind, we will employ Darwin's Diagram as a visual thread to guide the reader throughout the rest of this book—an illustration—with the proviso that the textual components necessary for fleshing out its deeper meaning can only be unpacked gradually as our book unfolds.

One central contention in the following pages is that Darwin clung, unwittingly perhaps, to what has been termed a *static worldview*. Considering that this notion will appear throughout the book, it seems appropriate at this point to clarify and offer a more formal definition of this expression. First, by "worldview" we mean a set or system of interconnected beliefs describing the overall features of the world. Second, by calling a worldview "static" we mean to denote a conception of a world that sees it changing only in minor, insignificant, ways over time. The essence of a static worldview lies in its stability: what can be observed around us today is pretty much what was found in the past and what will continue to exist in the future. The past, the present, and the future all fuse together to outline a world identical to itself, with nothing significantly new or different to be found in other time dimensions. Whatever explanations might be garnered to explain the creation of the world (whether by reference to God or to natural laws) it is argued that the world has remained roughly the same since its beginning, if such a beginning is postulated at all. On this view, for instance, the current order of our solar system, with the Sun at its center and planet Earth revolving on its third orbit, has always been thus. Similarly, living forms in their multifaceted shapes and behaviors have always populated our planet in their current form.

These somewhat abstract considerations will become more concrete to the reader as we gradually proceed with our analysis of the *Origin of Species*. At the outset, we would stress two main features of a worldview, as we will be using this notion here. First, it constitutes a coherent system of thought governing a wide range of entities, although these entities and the relations between them can be described in different ways. Second, it also represents a flexible system of thought that can be modified somewhat over time, allowing scientists to push its conceptual boundaries in new directions as their own understanding of the world develops. This is exactly what Darwin will do—or so we will argue in this book—ultimately venturing into an intellectual territory where the *static worldview* and the *evolutionary worldview* overlap.

The Present and the Past

The history of life dates back several billion years, with more complex forms appearing some 600 to 500 million years ago. Faced with such a long and deep history, one can intuitively grasp that any understanding of evolution must come in large part from studying the annals of life, that is, the actual fossil remains recorded in Earth's geological strata. Of course, forms currently alive are also products of evolutionary history, and as such carry important information about their past. No one, however, could justifiably claim that *all* relevant information about the history of life can be found among forms currently alive, especially considering the centrality of extinction to Earth's evolutionary history. To make such a claim would be equivalent to thinking one could learn about the Roman Empire by directing one's study to modern Europe. Of course, traces of the Roman Empire (in buildings, roads, and institutions, for example) can still be found, and we may even be able to draw tentative conclusions about aspects of Roman life from these artifacts despite the passage of time. Employing such a method, however, one can get no further than gathering together a hodgepodge of discrete facts. Without consulting the historical archives left to us in the form of ancient manuscripts or archeological vestiges, one has no principled way to connect them. This seemingly self-evident truth is obvious to us only because we take it for granted that history introduces changes over time, and see the present and the past as necessarily discontinuous. We are also accustomed to thinking that this gap widens over time, with more differences accumulating with each passing year, making the past less than full transparent in the present.

As surprising as it seems, Charles Darwin was only superficially committed to historical thinking. He knew, of course, that evolutionary changes would be introduced over geological time. Consciously or not, he continually downplays the importance of these changes, however. In a real sense, the explanatory foundation of the *Origin of Species* is an essentially ahistorical view. Let us see why.

First, rather than absorbing himself in the fossil record, Darwin expends significant effort discounting it, listing numerous reasons as to why it is uninformative (1859: 280, 287, 289, 293–301, 306). To this end, he presents a series of rationales for setting it aside:

1. Lack of significant fossil knowledge beyond Europe and the United States.
2. Unrecorded strata in the geological record.
3. Lack of correlation between the various geological formations.
4. Discrepancies in the rates of biological transformation and geological sedimentation.
5. A deficient fossilization process when seabeds rose above sea levels or when sediments do not accumulate.
6. Evolutionary continuity of lineages obscured by taxonomic debates.
7. Lack of preservation of soft body parts (i.e., soft body parts leave no fossil trace or sometimes only imprints).
8. Many fossil species are known only from a single specimen.

9. The pace of biological change is too slow to appear accurately in the geological record.
10. Gaps in the fossil record caused by migration of species.

Darwin goes well out of his way to raise these doubts, exhibiting a marked prejudice against the evidentiary status of the fossil record. Having significantly diminished the relevance of the past as a reliable source of information, and thus cut free from the empirical burdens they would impose, Darwin was essentially free to build his case primarily on currently living forms. At this point a distinction will prove useful, that between the area of biology devoted to studying past or extinct forms (paleontology) and the area concerned with investigating currently living or extant ones (neontology). Only two chapters out of thirteen in the *Origin* (excluding the final summary chapter) are exclusively concerned with paleontology, an asymmetry itself worth pondering.

Whatever his motivations, Darwin bases his biology on *neontology*, rather than on *paleontology*, as a number of modern evolutionary thinkers would now forcefully recommend. As one might expect, this approach to the past came with its share of limitations. Indeed, Darwin presents us with a theory based on the study of currently living forms, assuming the knowledge he has gathered applies equally to past evolutionary events with only small modifications. Our earlier analogy of studying modern Europe to learn about the Roman Empire, if anything, *understates* the limitations of this approach, since we are talking about eons, not centuries or even millennia, mere blinks of the eye in geological terms. The obstacles facing Darwin's approach to the evolutionary past are more or less the same: how much can actually be seen through the lens of neontology?

It is tempting to try to rehabilitate Darwin by claiming that he had no other choice, considering the poor state of the fossil record at time of the writing of the *Origin of Species* in the 1840s and 1850s, but to do so would be to neglect an important reality of the period: for several decades before the *Origin*, scholars had been gradually examining and trying to make sense of the fossil record, one that had begun to be exposed, slowly, from the late eighteenth century onward. Along with these discoveries, scientists of the day began to realize that Earth had a history both long and deep. While this thought is commonplace to us, it was a new idea at the time, one that many found hard to accept. Two main options were open for organizing the new information that arose out of this deep past: (1) to relate the fossil record to currently living forms, linking past forms to their modern-day equivalents; (2) to establish a succession of temporal cuts or slices/periods, each with its own set of fauna and flora, ever more different and strange as we delve deeper into geological time and move further away from extant forms. While Darwin chose the first option, scholars like Georges Cuvier and William Buckland became inspirational figures for the development of the second. Both options were very much on the table for scholars of the day, though the first would have implied a significant stretching of the conceptual envelope for anyone in the grips of a static worldview.

By choosing the first option, Darwin was following in the footsteps of someone who profoundly influenced his thinking, the geologist Charles Lyell, author of the *Principles of Geology* (1830–1833). Lyell tried to explain past geological events by appealing to geological processes seen to be operating in the present. Lyell was committed, as we now say, to *uniformitarianism*, the doctrine that states that nature is fundamentally consistent in character: the same laws and processes now in operation and directly observable in the present are valid for all geographical places and at all historical times. In order to study the past, then, the uniformitarian needs only to look at the present, since past and present phenomena are assumed to be continuous and thus roughly identical. Of course, such an assumption comes with the risk of overlooking past novelties finding no equivalents today. As early as 1832, the scientist-philosopher William Whewell openly recognized this danger. Assessing Lyell's approach to geology, Whewell wrote:

> It seems to us somewhat rash to suppose, as the uniformitarian does, that the information which we at present possess concerning the course of physical occurrences, affecting the earth and its inhabitants, is sufficient to enable us to construct classifications, which shall include all that is past under the categories of the present. Limited as our knowledge is in time, in space, in kind, it would be very wonderful if it should have suggested to us all the laws and causes by which the natural history of the globe ... is influenced—it would be strange, if it should not even have left us ignorant of some of the most important of the agents which, since the beginning of time, have been in action.[1]

It is worth noting that Charles Lyell was no fan of evolutionism. On the contrary, his uniformitarian stance was tailor-made for a world assumed to be, at the deepest level, homogenous and unchanging. To assume that the past and the present are equivalent is, indeed, to be committed to the belief that nothing significant has ever happened during the course of time, no matter how old planet Earth might be. It should be further noted that this belief comes in degrees, with a range of possible intermediary positions: the more one models the past on the present, the greater the risk of literally assimilating the two, and the smaller the role events of intervening years may play in one's explanatory framework. Of course, Charles Darwin did not embrace uniformitarianism as fully or as openly as Lyell had originally done. Darwin wrote the *Origin of Species* at a time when the intellectual pressure for recognizing important differences between the past and the present was increasing, and was quite ingenious and subtle in maintaining his commitment to the uniformitarian doctrine by adapting it to the context of a rising evolutionism. Let us review the multiple strategies Darwin uses; as we shall see, he has an interesting way of handling these cases.

When Extinct Forms Fall In-between Extant Forms

As more and more paleontological discoveries slowly began to reveal a deep past, Darwin was quick to *assimilate* potentially disruptive phenomena to currently living forms. Again, Darwin's approach insured that nothing significantly different could have existed in the past. A key passage of the *Origin* (1859: 303–6) is especially revelatory of his "assimilationist" approach. In the following examples, it should be noted how Darwin projects currently existing forms ever more deeply into the past (see Figure 10 for geological eras and periods):

1. To those who once declared that monkeys lived only during recent times (the Quaternary period), Darwin responds that they are now known to us from the oldest part of the Tertiary period.
2. He claims that whales are no longer confined to the Tertiary, extending also to the later portion of the Secondary era (Mesozoic).
3. He argues that in general, mammals can now be extended throughout most of the Secondary era (Mesozoic), whereas they were once restricted to the Quaternary and Tertiary periods.
4. He also holds that modern-looking crustaceans are now known from the Secondary era (Mesozoic), having been previously confined to the Tertiary.
5. Finally, he believes that a group of fish that includes the large majority of existing species—the Teleostean fish—can be traced back to the pre-Secondary era (Paleozoic).

In addition, to avoid the risk of seeing the past overflow into the present with novelties unknown in his day, Darwin uses the *principle of divergence* as a way to bracket all the past (*extinct forms*) within the biological diversity of currently living forms (*extant forms*). It is well known that Darwin relied heavily on the notion of "divergence" in order to describe what he thought was a significant

Era	Period
Cenozoic 66 million years ago	**Quaternary** **Tertiary**
Mesozoic 252 million years ago	**Cretaceous** **Jurassic** **Triassic**
Paleozoic 541 million years ago	**Permian** **Carboniferous** **Devonian** **Silurian** **Ordovician** **Cambrian**

Figure 10 Geological eras and periods. A temporal framework was created by geologists and paleontologists to organize our thinking about the history of life during the last 500 million years or a bit more.

reality of evolution, as reflected in his Diagram (see Figure 9). Furthermore, evolutionary divergence also meant that, for Darwin, related forms derive from a common ancestor. Darwin scholars have tended to overlook the fact, however, that the principle of divergence takes on a new meaning when seen against the background of Darwin's uniformitarian commitments. The following passage of the *Origin* is striking when viewed in this light:

> Let us now look to the mutual affinities of extinct and living species. They all fall into one grand natural system; and this fact is at once explained on the principle of descent. The more ancient any form is, the more, as a general rule, it differs

from living forms. *But ... all fossils can be classed either in still existing groups, or between them.* That the extinct forms of life help to fill up the wide intervals between existing genera, families, and orders, cannot be disputed. For if we confine our attention either to the living or to the extinct alone, the series is far less perfect than if we combine both into one general system. With respect to the Vertebrata, whole pages could be filled with striking illustrations ... showing how extinct animals fall in between existing groups. (Darwin 1859: 329) [our emphasis]

The worry about assimilating the past to the present was already a live concern in Darwin's time, having been raised by Whewell against Lyell, as in the quote above. In 2004, Marjorie Grene and David Depew assessed William Whewell's warning not to tackle the past armed only with current categories in the following words: "In light of mass extinctions spread across quite different geological eras, Whewell thought it decidedly unscientific for Lyell to shoehorn the history of life into categories cobbled up to fit contemporary life forms."[2] If the biological variability of all extinct forms ultimately falls within that observed among extant forms (as reflected in Darwin's Diagram), this would imply that the present time is more revealing of the fullness of the world than the past itself. To put it another way: as one moves further back in geological time, the biological variability one encounters among forms is continually reduced to the point of being confined to a presumed single common ancestor. In our attempt to grasp the whole history of life, the past is deficient relative to the present, since it only reaches its full expression in the latter. If this were so, it is tempting to ask why one should bother studying the past at all, since the present is always more informative. This is a fair overall description of the approach Darwin takes in the *Origin of Species*, one in large measure necessary for him to adopt, given his other presuppositions, in particular his commitment to the static worldview. He needed to ensure that nothing significant, evolutionarily speaking, would be hidden from us in the past.

Darwin's approach lacks a genuine historical dimension, an element more common to our thinking than his and characteristic of our own worldview. Were Darwin committed to a truly historical perspective, there would be absolutely no reason for him to minimize or overlook the existence of genuinely novel forms in the past. In order to press this criticism further, we need to contextualize it by comparing Darwin's general attitude regarding the past with that of more modern scholars, as found, for example, in the work of the American paleontologist Stephen Jay Gould (1941–2002). Gould argues that the biological variability ("disparity" in Gould's terminology) observed today cannot be squared with the variability seen in the fossil record. Consider, for instance, the animal kingdom as it now appears to us. It is divided into (twenty or thirty) different basic body plans, called *phylum*. These phyla include annelids (earthworms), chordates (vertebrates), corals, arthropods (insects), sponges, mollusks (clams), echinoderms (starfishes), among others. According to Gould, in the Cambrian period some 500 million years ago one finds somewhere between fifteen and twenty extinct body plans (phyla) whose body types are unlike anything now existing (see Figure 11).

Whether or not one accepts Gould's assessment in all its details, a number of modern scholars recognize that the diversity seen in ancient geological periods is not easily assimilated to extant forms, and have moved far away from a neontological approach and its blindness to evolutionary phenomena occurring earlier in the history of life. This is one point where Darwin's Diagram fails to reflect a modern view, as we have claimed above (see Figures 9 and 12). In fact, Darwin's Diagram cannot be understood as an accurate representation of a

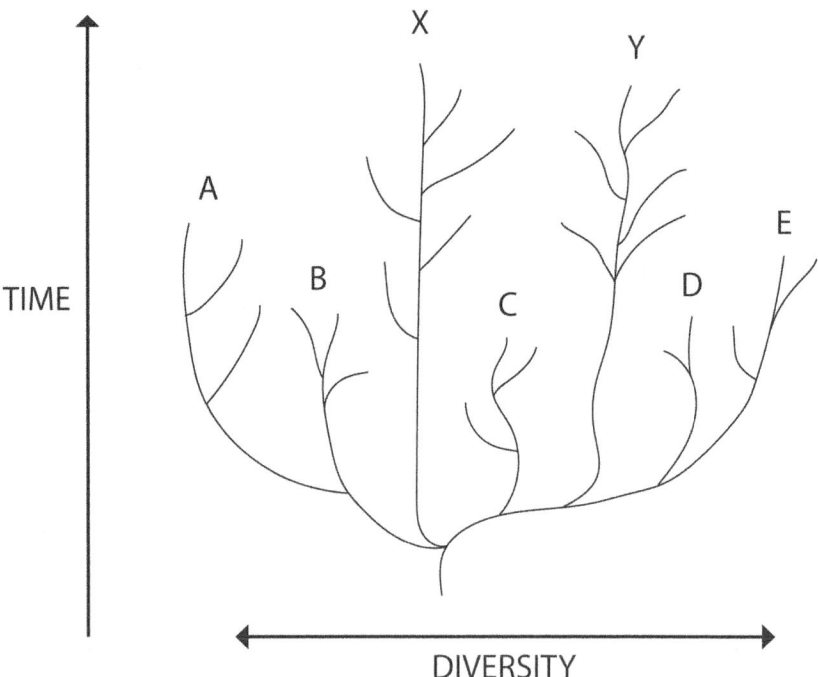

Figure 11 Decimation model. This view of life assumes that a significant amount of biological variability or disparity disappeared over the history of life (as seen in forms A, B, C, D, E), thus explaining why currently existing forms (X and Y) make for a very incomplete sample of that past's variability.

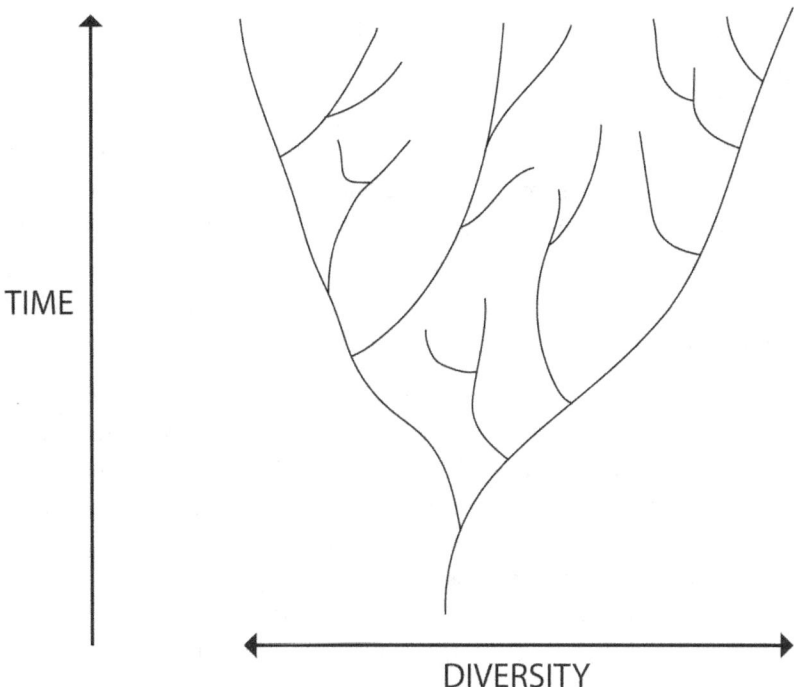

Figure 12 Cone of increasing diversity. This view of life adopted by Charles Darwin is based on the false assumption that extant forms register the most important manifestations of biological variability to have existed in the entirety of the history of life, with present time revealing what the past largely conceals. In the "cone of increasing diversity," it is believed that the past is entirely bracketed under currently existing forms, with no major forms having being lost in geological time.

portion of the "tree of life" at all; rather, it is merely a genealogy of extant forms. The two are quite different from one another. Far from attempting to trace the past by linking together pieces of the evidence then available, Darwin's Diagram *creates a past* by projecting the present backward in time. As Devin Griffiths rightly observes: "The sole illustration in the *Origin*, the famous branching tree of life, is an abstract of this *imagined past*; it is not an exemplary object but an ambiguous schema that traces a network of similarities and differences between examples that do not exist" [our emphasis].[3] A proper "tree of life" would involve a concerted effort to embrace the full historicity and complexity of the world, including the possibility of countless past novelties. Unfortunately, these are obscured in Darwin's presentation.

Recycling Variations by Projecting Them Backward in Time

Given what we have said thus far, it is unsurprising that throughout the *Origin of Species* we find descriptions of "common ancestors" constructed out of features selected from a menu of attributes possessed by extant forms, projected backward into geological times. At this juncture, we meet with one expression of Darwin's way of building arguments: the casual reader may think Darwin is referring to known common ancestors whereas, in reality, these are mere reconstructions by being imagined creatures. These putative "ancestors" ("unknown progenitors" in the text) can only be, unsurprisingly, pale echoes of forms found today:

> In the vertebrata, we see a series of internal vertebrae bearing certain processes and appendages; in the articulata, we see the body divided into a series of segments, bearing external appendages; and in the flowering plants, we see a series of successive spiral whorls of leaves. An *indefinite repetition of the same part or organ* is the common characteristic ... of all low or little-modified forms; therefore we may readily believe that the unknown progenitor of the vertebrata possessed many vertebrae; the unknown progenitor of the articulata, many segments; and the unknown progenitor of flowering plants, many spiral whorls of leaves. (Darwin 1859: 437) [our emphasis]

The expression "indefinite repetition of the same part" in this passage seems to suggest a kind of recycling process at work, favoring repetition at the expense of novelty. Of course, Darwin recognizes that evolutionary change occurred with the passage of time. But his entirely hypothetical "ancestors" are little more than dim reflections, mere shadows of current forms. Here again it is valuable to situate Darwin within the thinking of his time. In this wider context, one sees the kinship of his explanatory approach with the widely held *preformationist* doctrine of the eighteenth century and before. The guiding thought underlying this naturally appealing view of life is that any form originates from a much smaller version of itself: all the main features of an organism are already in place at the earliest stage of its conception; its growth is not accompanied by the emergence of new features, since the process only involves an increase in size of features already existing from the beginning. The reasoning here is that whatever we can now observe originated from an earlier, less fully worked-out version of itself. Transferred to the case of evolution, the starting point is once again individuals existing today. Antecedent forms, the "common ancestors," are imagined as cruder versions of their current selves, needing only time for their full deployment. Under his preformationist stance, Darwin's postulated progenitors or common ancestors that look strangely familiar. It should be noted that the doctrine of *preformationism* is yet another manifestation of the system of thought described earlier in this chapter, the *static worldview*.

Scientists have since been largely able to extricate themselves from preformationist views and have let go of the thought that the evolution of a species can be roughly modeled on the development of an embryo, as preformationist thinking would commend. In contrast to Darwin, biologists today recognize features in classes or phyla of the past that go well beyond today's variability. This provides explanatory room for the recognition of genuine evolutionary novelties. In the vertebrates, for instance, bone or cartilage vertebrae that make up the vertebral column are traced back not to smaller vertebrae, but to ever more primitive structures: an axial rod made of fibrous connective tissue wall around cells called *a notochord*, itself possibly derived from a proto-notochord state, a structure that may ultimately be traceable to a mere dorsal nerve cord.

The Taxonomic Position of the Main Forms Has Remained Fixed Throughout All of Evolutionary History

Another of Darwin's commitments reveals his great distance from modern scholars, in particular his reluctance to accept the full openness of evolution, a cornerstone of modern evolutionary thinking as exemplified by a number of contemporary theories (see our Chapters 1 and 2). Darwin claimed that no matter how much evolutionary change was introduced during the course of time, the taxonomic position of the main groups currently alive relative to one another remained the same. His view implies that evolutionary change was so insignificant throughout the entire geological record of life that the taxonomic order has remained fixed or stable, undisturbed by the hazards and contingencies associated with the critical need for adaptation. This is more or less a recapitulation of the claim that what neontology tells us applies equally to paleontology. Darwin's ahistorical approach, in effect, puts evolution in a straitjacket, insofar as it registers a conviction that nothing significantly new was to be expected from future paleontological discoveries. In order to get a feel for the texture of Darwin's thinking, the reader is invited to read his words in full. Explaining his Diagram (see Figure 9), he writes:

> The reader will best understand what is meant, if he will take the trouble of referring to the diagram ... We will suppose the letters A to L to represent

allied genera, which lived during the Silurian epoch, and these have descended from a species which existed at an unknown anterior period. Species of three of these genera (A, F, and I) have transmitted modified descendants to the present day, represented by the fifteen genera (a^{14} to z^{14}) on the uppermost horizontal line. Now all these modified descendants from a single species, are represented as related in blood or descent to the same degree ... yet they differ widely and in different degrees from each other. The forms descended from A, now broken up into two or three families, constitute a distinct order from those descended from I, also broken up into two families. Nor can the existing species, descended from A, be ranked in the same genus with the parent A; or those from I, with the parent I. But the existing genus F^{14} may be supposed to have been but slightly modified; and it will then rank with the parent-genus F; just as some few still living organic beings belong to Silurian genera. So that the amount or value of the differences between organic beings all related to each other in the same degree of blood, has come to be widely different. *Nevertheless their genealogical arrangement remains strictly true, not only at the present time, but at each successive period of descent.* All the modified descendants from A will have inherited something in common from their common parent, as will all the descendants from I ... All the descendants of the genus F ... are supposed to have been but little modified, and they yet form a single genus. But this genus, though much isolated, will still occupy its proper intermediate position; for F originally was intermediate in character between A and I, and the several genera descended from these two genera [A and I] will have inherited to a certain extent their characters. (Darwin 1859: 420–2; see also 1872: 369–70) [our emphasis]

For Darwin, then, the relative positions of lineages A, F, and I (including all their descendants) have remained the same since the earliest fossils, with lineage F in the middle, A on the left side, and I on the right side.

At this stage of our investigation, the reader may be becoming suspicious about something we ourselves have only slowly come to realize about the *Origin of Species*: how little room Darwin leaves for evolutionary novelties and the extent to which he contributes to expanding the conceptual envelope of the static worldview (with further expressions of this tendency to be presented below). Geological or historical time, for him, cannot be truly creative, since the present is believed to have registered everything essential about the past and is indeed more informative than the past. As suggested above, Darwin implicitly assumes that the fullness of the world is best revealed by extant forms and can be fully appreciated by examining them directly, past forms being inferior versions of present ones.

Rooting Evolution in Two Embryological Laws

As we have seen above, two corollary implications of Darwin's view made it hard for him to embrace anything approaching modern evolutionism, as they run directly counter to it: (1) little explanatory room is left to accommodate for the rise of genuine evolutionary novelties during the course of time; (2) evolution is ensconced in a straitjacket of unchanging taxonomic arrangements maintained since the beginning of life. These two commitments will come into sharper focus as we review Darwin's approach to embryology.

By the first half of the nineteenth century, biologists had realized that the field of embryology—the study of embryological development in different species—revealed that some species shared similarities during their earliest stages of development not visible in later stages. The comparison of the embryological development in mammals, reptiles, and fish, for instance, showed such similarities. Two rival explanations attempted to account for this fact:

1. *The law of parallelism.* Scholars such as Johann Meckel, Étienne Serres, and Ernst Haeckel envisioned embryological development as a linear sequence of stages organized along a hierarchy heavily oriented toward humans, and for which embryos of all species unfold along an identical path, with lower animals such as fish stopping short of the higher stages with higher animals climbing further upward in level. Under this hypothesis, humans are assumed to be the most complex among higher animals, which explains why they alone reach the highest embryological stage. The law is said to be "parallel" precisely because all the species are believed to follow exactly the same embryological path, differing only in how far along they have travelled on this path.
2. *The law of specialization.* Karl Ernst von Baer proposed an alternative explanation, arguing instead that vertebrate embryos of all species start their development from a similar (not necessarily identical) stage, only to take off in separate directions as each species acquires its unique features and specializations. This is seen today, for instance, in the different forms belonging to separate classes such as fish, reptiles, and mammals. This alternative to "parallelism" focuses on the "divergent" embryological paths taken by separate species.

Although the nineteenth-century biologists who attempted to account for embryological phenomena were not all committed to the evolutionist doctrine, it is easy to see why such explanations strongly appealed to evolutionists.

Whether one accepts the parallel or the specialization hypothesis, it can easily be argued that embryological similarities observed between species point to some sort of genealogical or phylogenetic connections between them. In other words, if fish, reptiles, and mammals show similarities in their respective early embryological developments, it is because they are bound together in common ancestry. These similarities point toward a common ancestor—which has in effect passed on the "raw material" out of which differentiated species grew. Although Darwin scholars do not agree on which of the two hypotheses Charles Darwin subscribed to—and setting aside the fact that the specialization account seems more congenial to Darwin's commitment to evolutionary divergence— Darwin understandably relied on the field of embryology to promote the notion of "common ancestry."

With this in mind, we should look more closely at precisely why Darwin thought embryology might be successfully employed to support the notion of common ancestry. It is at this precise explanatory junction that we see Darwin again favoring the more conservative, constraining option, the one most tightly limiting evolutionary flexibility. As we have seen, Darwin argues that evolutionary divergence among related species is the product of their adaptation to different conditions. Adaptive pressure is one key reason that closely related species will evolve in different directions (divergence), each adopting different strategies for survival. Now, if adaptive pressure modifies species, we would expect that their embryological development will be modified as well, thus concealing, or even erasing, their genealogical connections. If this is the case, embryology cannot be a reliable guide, since the embryo is equally susceptible to evolutionary parallelism and evolutionary convergence. As discussed in earlier chapters, under the species' struggle to find new adaptive solutions, natural selection may lead unrelated life forms to come up with similar solutions to similar challenges, as we saw in our earlier example of sharks and dolphins (see our Chapter 1): they look alike in their overall elongated hydrodynamic shape *not* because they are closely related, but because they have adapted to the same conditions (in this case, the high-density medium of water).

Darwin (1859: 86–7, 440) recognizes that natural selection may well modify the embryological development of species to the point of blurring the phylogenetic connections. Darwin, however, was not yet prepared to relinquish the potential argumentative assets provided to him by the field of embryology. For him, not only can embryology offer key information about the common ancestors of presumably related species, it may also allow us a peek at the actual shape of such ancestors. In the sixth edition of the *Origin of Species*, the reader encounters this unambiguous passage: "As the embryo often shows us more or less plainly the structure of the less modified and ancient progenitor of the group, we can see why ancient and extinct forms so often resemble in their adult state the embryos of existing species of the same class" (Darwin, 1872: 396). Why, one might ask at this point, did Darwin believe he could so easily sidestep the potential problems associated with embryology, once evolutionary parallelism and convergence are taken into account? The answer is to be found in three complementary theoretical commitments Darwin held with respect to embryological development (Darwin, 1859: 439–50). Darwin thought the following considerations limited the potential threat to his use of embryology:

1. *Natural selection has little impact on embryos when shielded in the mother's womb, or on young children when protected and fed by the parents, since they don't have to provide for themselves at such an early stage of their life.*

On the basis of this first commitment, Darwin subscribes to an embryological theory founded on two additional explanatory components:

2. *New features or changes in the embryological development of species appear during later stages rather than earlier stages.* This claim is meant to capture the notion that natural selection will not impact development until later embryological stages, thus leaving the earlier stages pristine. If this is the case, embryology can be used to gather information about common ancestors.

3. *New features or changes in embryological development of species are manifested at the same age throughout all generations of the same lineage.* This new claim reinforces the notion of a deep stability in embryological development, one that even allows the use of changes or the introduction of adaptive features as a means of tracing genealogy, considering that they occur in a clockwork fashion in all generations. If this is the case, changes brought by the impact of natural selection on embryological or ontogenetic development can still be useful for investigating phylogenetic relationships.

Darwin took great pains to ensure that embryology remained a key explanatory apparatus incorporated in his general theory. What is most remarkable is that Darwin (1859: 446, 449) acknowledged that the last two conditions of his embryological theory were as yet unproven, while hoping they would be in the future. Moreover, the reader of Darwin's embryological section encounters counterfactual evidence Darwin himself recognizes (see especially

Darwin: 1859: 444). Why, then, would Darwin so decisively take sides in favor of a yet unproven embryological theory?

The answer lies, we suggest, in Darwin's understanding of the past, which, though unremarkable given the static worldview that was not uncommon at that time, is quite peculiar when set alongside our modern view. First, his embryological theory deprived evolution of the required flexibility for life to adapt itself fully to its challenges. His supposition that embryological changes occur in a clockwork fashion over the history of a lineage is a clear expression of this rigidity. Second, his embryological theory reveals a strong tendency toward immutability. By stating that earlier embryological stages remained pristine (in contradistinction to later stages), Darwin anchored related lineages in ancient common ancestors that had not, in his view, undergone significant modification as compared to their modern-day counterparts. Darwin is convinced that extant forms give him a clear window into the past—their embryological developments being one such window—irrespective of the numerous chance events that are today judged to play such a central role in evolution: profound modifications under adaptation, evolutionary parallelism and convergence, and mass extinctions. Running in precisely the opposite direction, Darwin's embryological theory is a time machine that denies evolution's creativity and novelty.

A Modern-Looking World as Far Back as We Can Probe Geologically

As we have seen, Darwin's view is largely ahistorical and unidimensional, heavily weighting the present. It is therefore unsurprising that in the *Origin of Species* we encounter a world that is, in its essentials, pretty modern-looking as far back as we can survey. In the sixth edition of the *Origin of Species*, one reads: "As all the living forms of life are the lineal descendants of those which lived long before the Cambrian epoch, we may feel certain that the ordinary succession by generation has never once been broken, and that no cataclysm has desolated the whole world" (1872: 428). At a superficial level, this seems merely to suggest that the process of life on Earth has never experienced any serious disruptions since its beginnings. When seen alongside other sections of the *Origin*, however, these words reveal a rather different meaning. To see why this is so, we will need to dig deeper into these sections.

Before working our way through these various parts, let us lay out several assumptions regarding Darwin's approach to the past. For Darwin, the history of life is not organized around a *vertical* succession of independent groups of forms, each living in its own distinct geological period; rather, life is organized around a *horizontal* distribution of classes seen in extant forms (such as fish, reptiles, and mammals), each projected backward in time. This is exactly what is illustrated in Darwin's Diagram when applied to a single class (see Figure 9). In order to get a grip on Darwin's overall view of the history of life, one must lay a series of these diagrams side by side (one class = one diagram; see Figure 13). What emerges from this exercise is a picture of the independent and parallel evolution of as many different classes as are known today. Because Darwin imagines the past of each class by projecting the variability known in its extant representatives backward in time, he only manages to envision classes that are both separate from one another and modern-looking in their conformation, despite his use of embryology as a tool to attempt to probe more deeply:

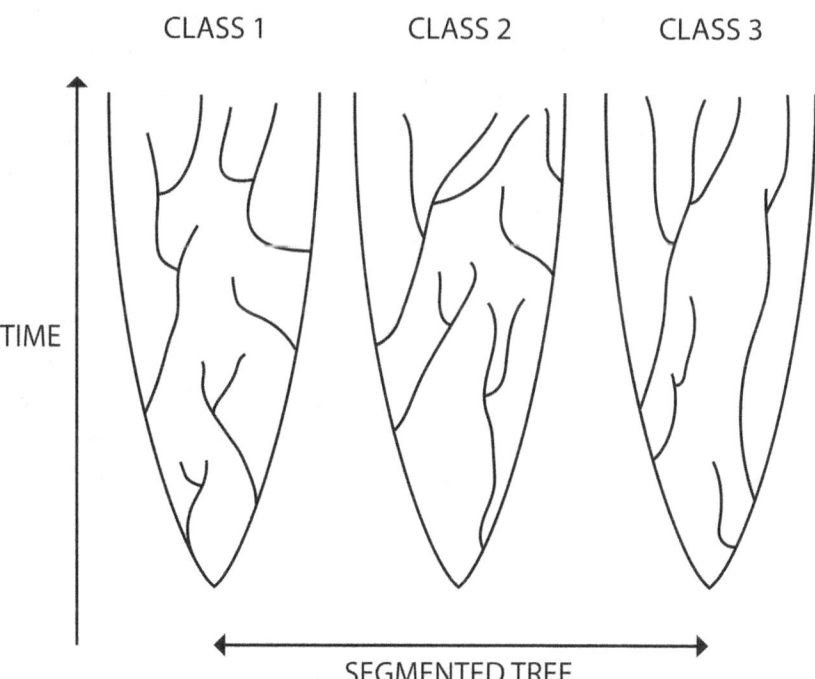

Figure 13 Darwin's overall view of the history of life. In order to grasp what Charles Darwin had in mind when imagining life on Earth as a whole, one must reconstruct it by laying a series of Diagrams side by side (one class = one diagram). What emerges from this exercise is the independent evolution of as many different classes as are known today, as seen in a segmented tree of life.

> Thus the embryo comes to be left as a sort of picture, preserved by nature, of the former and less modified condition of the species. This view may be true, and yet it may never be capable of full proof. Seeing, for instance, that the oldest known mammals, reptiles, and fishes strictly belong to their own proper classes, though some of these old forms are in a slight degree less distinct from each other than are the typical members of the same groups at the present day, it would be vain to look for animals having the common embryological character of the Vertebrata, until beds rich in fossils are discovered far beneath the lowest Cambrian strata—a discovery of which the chance is small. (Darwin 1872: 310)

To get a more complete picture of Darwin's view, other of his assumptions must also be brought to light. First, Darwin holds that the development of the various independent classes was an extremely slow process requiring a significant amount of time since the beginnings of life. When explaining his Diagram (1859: 420–2), Darwin argues that a class developed in the following way: from distinct allied genera in the Silurian period (more than 400 million years ago in today's standard), to distinct families, and eventually to distinct orders as seen today. In addition, Darwin tells us that the length of time separating the Silurian from today represents about half of the entire time needed for life to unfold, meaning that the original form at the root of a class in the Diagram required a huge amount of time even to move from one original species in the pre-Silurian period to several allied genera in the Silurian.

The modern reader can readily gather that from these considerations Darwin's sense of history was quite different from ours. Whereas today's evolutionists see the history of life as full of chance events and novelties—replacements of major taxonomic groups, mass extinctions, slow or rapid turnovers—Darwin's world is one of an extremely slow and gradual development of what will eventually become extant classes, entities that have existed since life's very beginnings. In Darwin's world, nothing truly new ever occurs at the higher taxonomic levels (i.e., classes, phyla). Indeed, Darwin holds that the classes we know today may have already been in existence since a very long time: "[A]t the most remote geological period, the earth may have been as well-populated with many species of many genera, families, orders, and classes, as at the present day" (1859: 126). This statement fits perfectly with the previous quotation, in which Darwin explains that the classes of the oldest known fish, reptiles, and mammals were already quite distinct from one another at that time. No surprise, then, that a close look at the *Origin of Species* leaves the reader with a sense of a world that was fully constituted long ago, a nearly static and eternal world in all its essential features. Darwin's Silurian or Cambrian world looks strangely like ours today. For him, groups at high taxonomic levels are entities profoundly entrenched in nature, ever more dominant, driving others to extinction (a notion we shall return to in our Chapter 6).

The ongoing evolutionary dynamic of ever-more dominant groups replacing less successful ones has been going on since very ancient times, recycling the same variations seen today within the confines of permanent and independent classes, according to Darwin. Apparently, the domination and expansion of today's classes will likely persist for a considerable amount of geological time in the future. Indeed, looking back at the long history of life since the pre-Silurian or pre-Cambrian period, Darwin projects himself into the future: "Hence we may look with some confidence to a secure future of equally inappreciable length" (1859: 489; see also 1872: 428). From Darwin's Diagram, one gathers that classes are continuing their expansion, as seen in the ongoing divergence process of the most recent forms. Here, we encounter the ultimate explanatory function of the *principle of divergence* in Darwin: to bracket the entire evolutionary history of a class. When the time dimension of the divergence process (expansion) is reversed, one gets convergence backward in time, pointing toward a presumed common ancestor.

By using modern biological types as the basis for recreating past forms, Darwin commits himself to the notion that what we see of the modern world has in fact been in place for a very long time. Darwin comes up against a sort of Silurian or Cambrian barrier: before the barrier, nothing is known about life, and he admits only of hypothesis and speculation; after the barrier, life is largely organized around familiar, well-defined groups as seen in extant forms; at the barrier itself, life seems to spring out of nothing, suddenly, but more or less fully formed. This is illustrated in the following quote (this time referring to the example of the invertebrates):

> [N]umber of species of the same group, suddenly appear in the lowest known fossiliferous rocks. Most of the arguments, which have convinced me that all the existing species of the same group, have descended from one progenitor, apply with nearly equal force to the earliest known species. For instance, I cannot doubt that all the Silurian trilobites have descended from some one crustacean, which must have lived long before the Silurian age, and which probably differed greatly from any known animal. Some of the most ancient Silurian animals, as the Nautilus, Lingula, &c., do not differ much from living species; and it cannot

on my theory be supposed, that these old species were the progenitors of all the species of the orders to which they belong, for they do not present characters in any degree intermediate between them. (Darwin 1859: 306)

No matter how much Darwin may have speculated about life originating from a limited number of truly primitive ancestors (1859: 484, 488, 490), his biology is designed to deal with modern-looking forms only, leaving little room for evolutionary change and novelties. Moreover, while historians of science have been fond of contrasting Darwin's so-called profound commitment to "evolutionism" with the views of much more moderate contemporaries such as Thomas H. Huxley and Richard Owen, the fact remains that they all came to quite similar understandings in the 1860s. For instance, Huxley writes in 1862:

[E]nough has been said to justify the statement that, in view of the immense diversity of known animal and vegetable forms, and the enormous lapse time indicated by the accumulation of fossiliferous strata, the only circumstance to be wondered at is, not that changes of life, as exhibited by positive evidence, have been so great, but that they have been so small.[4]

What struck Huxley, above all, was the extent to which extant forms at high taxonomic levels remained similar through time. When speaking of plant orders, he added: "There are two hundred known orders of plants; of these not one is certainly known to exist exclusively in the fossil state. The whole lapse of geological time has as yet yielded not a single new ordinal type of vegetable structure."[5] Richard Owen came to a similar conclusion with respect to invertebrate classes in 1860:

But as to the larger groups of ... true invertebrate animals, it may be affirmed that every known fossil belongs to someone or other of the existing classes, and that the organic remains of the most ancient fossiliferous strata do not indicate or suggest that any earlier and different class of being remains to be discovered, or has been irretrievably lost.[6]

In the 1860s, Darwin was hardly alone in promoting a biology deeply committed to the study of modern forms, as if geological times had revealed a past for all intents and purposes equivalent to the present. This tradition in biology, rooted in the static worldview, adapted itself to the rise of the evolutionist doctrine more by avoiding its implications than by confronting them.

Conclusion

Few modern scholars have remarked on the influence of the static worldview on the *Origin of Species*. One notable exception is Camille Limoges, who, as long ago as 1972, caught a glimpse of Darwin's ahistorical approach:

> As always during the eighteenth century, recourse to the notion of origin serves the purpose of thinking in all clearness about what would still exist today and under the very same mode but for a different state, *more opaque* to our representation ... The *Origin of Species* was in no way a treatise about origins: the features of the economy of nature were not founded on any such question; on the contrary, transmutation of species finds its rules in contemporaneity, or rather in atemporality, that is, the synchronic ecological operation of natural selection.[7] [our emphasis and translation from French]

Although Limoges does not register the extent to which Darwin was still in debt to the static worldview, he aptly describes the notion of a past reflecting the present as a sort of mirror image, one that, as a mere reflection of the present, is more diffuse and somewhat hazy. In so doing, Limoges is explicitly drawing upon Michel Foucault's analysis in *The Order of Things*, originally published in 1966. Speaking of the Classical Age of the eighteenth century, Foucault had this to say:

> In the eighteenth century, to return to the origin was to place oneself once more as near as possible to the mere duplication of representation ... The order of nature was conceived, prior to any catastrophe, as a table in which beings followed one another in so tightly knit an order, and upon so continuous a fabric, that in going from one point of this succession to another one would have moved within a *quasi-identity*, and in going from one extremity of it to the other one would have been led by the smooth expanse of "likeness."[8] [our emphasis]

By not engaging seriously with the past as a reality of its own, Darwin condemned himself to taking on board much of the prevailing eighteenth-century thinking about time and history in general. Foucault, like many today, assumes that Darwin succeeded in transcending the Classical representation of time, embracing its modern form.[9] We argue, on the contrary, that Darwin was, unwittingly perhaps, a standard-bearer of this understanding of time, carrying it in the nineteenth century.

Annotated Bibliography

The two editions of the *Origin of Species* referred to in this chapter are Darwin, C. (1859), *On the Origin of Species,* London: John Murray; Darwin, C. (1872), *The Origin of Species,* 6th edn, London: John Murray.

Charles Darwin's Diagram is found in the middle of Chapter 4 in all six editions of the *Origin of Species.*

William Whewell warns of the risk of studying the past from the perspective of the uniformitarian doctrine in Whewell, W. (1832), "[Review of volume 2 of Lyell's] Principles of Geology," *Quarterly Review*, 47: 103–32.

Marjorie Grene and David Depew's assessment of William Whewell's warning not to tackle the past with current categories is found in Grene, M. and D. Depew (2004), *The Philosophy of Biology: An Episodic History*, Cambridge: Cambridge University Press.

For a clear and concise presentation of the main issues involved in the debate concerning the Burgess Shale animals of the Cambrian period, see Sterelny, K. (2007), *Dawkins vs. Gould: Survival of the Fittest,* 2nd edn, Cambridge: Icon Books, Chapter 10. For arguments to the effect that the disparity or diversity in ancient geological periods cannot easily be assimilated to extant forms, consult Gould, S. J. (1989), *Wonderful Life: The Burgess Shale and the Nature of History*, New York: W. W. Norton; Gould, S. J. (1991), "The Disparity of the Burgess Shale Arthropod Fauna and the Limits of Cladistic Analysis: Why We Must Strive to Quantify Morphospace," *Paleobiology*, 17: 411–23; Seilacher, A. (1992), "Vendobionta and Psammocorallia: Lost Constructions of Precambrian Evolution," *Journal of the Geological Society* (London), 149: 607–13; Gee, H. (1996), *Before the Backbone: Views on the Origin of the Vertebrates*, London: Chapman and Hall, 295–9; Conway Morris, S. (1998), *The Crucible of Creation: The Burgess Shale and the Rise of Animals*, Oxford: Oxford University Press, 12–14, 27–30, 169–218.

On the origin of vertebrates, consult the following literature: Gee, H. (1996), *Before the Backbone: Views on the Origin of the Vertebrates*, London: Chapman and Hall; Saxena, R. and S. Saxena (2008), *Comparative Anatomy of Vertebrates*, UK: Anshan, Chapter 1; Kardong, K. (2012), *Vertebrates: Comparative Anatomy, Function, Evolution*, 6th edn, New York: McGraw-Hill, Chapters 2–3.

For more about the doctrine of preformationism, consult Roger, J. (1998), *The Life Sciences in Eighteenth-Century French Thought*, Stanford: Stanford University Press.

Darwin scholars disagree about which embryological thesis Darwin subscribed to (i.e., the law of parallelism or the law of specialization). Interested readers may consult Gould, S. J. (1977), *Ontogeny and Phylogeny*, Cambridge: Belknap Press, 69–74; Ospovat, D. (1981), *The Development of Darwin's Theory: Natural History, Natural Theology, and Natural Selection, 1838-1859*, Cambridge: Cambridge University Press, 146–69; Richards, R. J. (1992), *The Meaning of Evolution*, Chicago: University of Chicago Press; Nyhart, L. K. (2009), "Embryology and Morphology," in M. Ruse and R. J. Richards (eds), *The Cambridge Companion to the "Origin of Species,"* 194–215, Cambridge: Cambridge University Press.

For Charles Darwin, classes and higher taxonomic entities avoided extinction altogether. Darwin's position can be assessed by combining two sources of information: (1) the principle of divergence ensures that all past forms belonging to different classes fall in-between extant ones, a principle applied since the beginning of life (1859: 329; 1872: 301); (2) separate groups belonging to distinct classes already existed in the oldest known strata (1859: 126; 1872: 97). In contrast to Darwin, scholars today recognize the extinction of several classes and phyla, for instance, of invertebrate classes like *Trilobita* and *Blastoidea*, of vertebrate classes like *Placodermi* and *Acanthodii*, of an animal phylum like *Vetulicolia*, and of other possible phyla part of the Ediacaran and Burgess Shale faunas. It is true that Darwin recognized the extinction of trilobites and ammonites, but for him, these groups were either of the family or order level only (1859: 321; 1872: 297).

No matter how much Darwin's view is sometimes contrasted with those of his contemporaries like Thomas H. Huxley and Richard Owen in the secondary literature, we hold that the three shared quite similar unmodern evolutionary views about the fossil record. Readers wishing to consider primary sources may consult Huxley, T. H. (1862), "The Anniversary Address," *Quarterly Journal of the Geological Society of London*, 18: xl–liv; Owen, R. (1860), *Palaeontology or a Systematic Summary of Extinct Animals and Their Geological Relations*, Edinburgh: Adam and Charles Black.

For the representation of the Classical Age and its view of the past as a merely obscured or diminished state as compared to the present, see Foucault, M. (2002), *The Order of Things: An Archaeology of the Human Sciences*, London: Routledge Classics. Although Camille Limoges may not have fully realized it himself, we think he managed to put his finger on this same phenomenon in Charles Darwin; see Limoges, C. (1972), "Introduction," in C. Linné, *L'équilibre de la nature*, 7–22, Paris: J. Vrin. For an analysis of Foucault's views on this matter, see Gutting, G. (1989), *Michel Foucault's Archaeology of Scientific Reason*, Cambridge: Cambridge University Press, Chapters 4 and 5.

Notes

1. Whewell, W. (1832), "[Review of volume 2 of Lyell's] Principles of Geology," *Quarterly Review*, 47: 126.
2. Grene, M. and D. Depew (2004), *The Philosophy of Biology: An Episodic History*, Cambridge: Cambridge University Press, 171.
3. Griffiths, D. (2016), *The Age of Analogy: Science and Literature between the Darwins*, Baltimore: Johns Hopkins University Press, 11.
4. Huxley, T. H. (1862), "The Anniversary Address," *Quarterly Journal of the Geological Society of London*, 18: l.
5. Huxley, T. H. (1862), "The Anniversary Address," *Quarterly Journal of the Geological Society of London*, 18: xlviii.
6. Owen, R. (1860), *Palaeontology or a Systematic Summary of Extinct Animals and Their Geological Relations*, Edinburgh: Adam and Charles Black, 18.
7. Limoges, C. (1972), "Introduction," in C. Linné (ed), *L'équilibre de la nature*, Paris: J. Vrin, 12.
8. Foucault, M. ([1966] 2002), *The Order of Things: An Archaeology of the Human Sciences*, London: Routledge Classics, 358.
9. Gutting, G. (1989), *Michel Foucault's Archaeology of Scientific Reason*, Cambridge: Cambridge University Press, 190–3.

4

Divergence: A Geometry That Shatters Creative Time and Novelty

As we have just seen in Chapter 3, a cursory look at Charles Darwin's famous Diagram (see Figure 9) may easily leave the reader with the impression that evolution is haphazard, with lineages branching out in all sorts of evolutionary directions. It would thus seem to portray exactly what we should expect of a modern view: a graphic illustration of *evolutionary contingency*, a cornerstone of many evolutionary views currently in circulation. However, there is great value in looking at the roots, stems, branches, and twigs found in Darwin's Diagram. Here, as elsewhere, we are recommending reading Darwin with an awareness that things may not always be as they seem or as we have been taught. The Diagram is a case in point: a slightly closer inspection reveals something rather strange. If one traces the path taken by each tiny evolutionary line belonging to the two main diversifying lineages in time (A and I), one realizes that roughly two-thirds of the lines that manage to make it to the next geological horizon (lines I, or II, or III, for example) are the most divergent of the forms; less divergent ones are consigned to the wastebasket of evolutionary history.

Take, for example, the oldest common ancestor of lineage A on the bottom left side of the diagram. Of the six tiny lines found there, only the two most divergent ones—the one branching off to the extreme left and the one to the extreme right—proceed to the next geological horizon (I). Each then becomes a new starting point for later forms evolving toward the next geological horizon (II). This pattern is repeated throughout the diagram: the more these tiny branches diverge, the more likely they are to persist. To see the overall predominance of this pattern, begin from any of the following four reference points: the most recent descendant of lineage A in geological horizon XIV on the far upper left side (a^{14}), the most recent descendant of lineage I in geological horizon XIV on the far upper right (z^{14}), and their two respective common ancestors (A and I) in geological horizon 0. What this tells us is that, for Darwin, all past and extinct forms fall in-between currently existing ones, which is precisely the case we have been making thus far regarding the predominance of divergence in his thinking: for Darwin, the history of life is organized around a giant "V."

This might seem like a small matter, a technical oversight or oversimplification on Darwin's part, adopted only for the purposes of illustration. However, the suggestion that divergent forms drive others to extinction is sharply at odds with the principle of evolutionary contingency deemed an essential feature of evolution by many today. Instead of seeing life forms ramifying freely in all directions through an open-ended evolutionary process, in the Diagram we see them channeled into a preestablished pattern. This points to something *rigid*, *geometrical*, and *predictable* in Darwin's view of evolution. This feature of his Diagram suggests a closer examination is in order, particularly at Darwin's textual explanations in the *Origin*, when interpreting this diagram.

Divergence and Specialization

Darwin's preference for a patterned approach has a fairly clear rationale. For him, evolutionary change is a direct consequence of a world filled to capacity with life forms, meaning organisms are continually brushing up against each other and encroaching upon each other's territory. Darwin's world is characterized by overpopulation, a state of affairs that inevitably leads to constant, harsh competition among forms, who engage in life-or-death struggles to occupy the limited number of available ecological niches or places in the economy of nature. Darwin argues that the royal road for survival in this ruthless environment is the path that steers them clear of other competitors. To put it another way: what separates winners and losers in evolutionary terms is their ability to go their own way and carve out new strategies for meeting basic needs, outcompeting their rivals by avoiding actual and potential competitors. It is at this juncture that we reach the heart of Darwin's reasoning.

For Darwin, successful lineages do not change in a random fashion, evolving in every possible direction, following in random measure the pathways of reticulate evolution, parallel evolution, divergent evolution, and convergent evolution. Instead, successful species are largely those that literally "diverge" from their competitors. As we have seen, Darwin believes significant divergence spells success in the evolutionary game, with degree of divergence directly correlated with it. Darwin is so committed to this idea that he awards it an explanatory status approaching a *law-like* manifestation. He does so under the name of a "general rule," an evolutionary outcome that is fairly predictable, in a way that would be deemed excessive from modern perspectives grounded in evolutionary contingency. As is often the case in the *Origin*, Darwin's argument is presented in a convoluted way: starting with the suggestion that *persistent life forms are not invariably the most divergent ones*, Darwin moves into the heart of the matter by recognizing that this phenomenon is, actually, so common that its applicability should be subsumed under a *general rule*:

> I am far from thinking that the *most divergent varieties will invariably prevail and multiply*: a medium form may often long endure, and may or may not produce more than one modified descendant; for natural selection will always act according to the nature of the places which are either unoccupied or not perfectly occupied by other beings; and this will depend on infinitely complex relations. *But as a general rule, the more diversified in structure the descendants from any one species can be rendered, the more places they will be enabled to seize on, and the more their modified progeny will be increased.* (Darwin 1859: 118–19) [our emphasis]

In other words, whereas the most divergent forms are not *always* the ones positively selected, this is *very often* the case. It is also worth noting that, for Darwin, "divergence" and "specialization" are closely related notions (1859: 112–16). Among species in a potential state of competition with one another, evolutionary opportunities are created when potential competitors avoid confrontations by specializing, by developing unique talents, so to speak. For instance, carnivorous quadrupeds became successful by adopting a range of unique survival strategies: (1) feeding on different kinds of foods, some becoming less carnivorous than others; (2) inhabiting new ecological settings with some forms climbing trees and others taking to aquatic habitats (lakes and streams). The adaptive incentive to evolve forward is created when divergence is accompanied by adaptive specialization. Darwin argues that a greater number of different life forms can remain in the same natural milieu only if each manages to carve out its own unique and specialized place within it. To put it another way, competition is reduced between forms living in the same ecological zone when each exploits a different set of resources, thus allowing for the survival of a greater number of them in that area.

The Competitive Exclusion Principle

Ultimately, Darwin's conviction that evolution is dominated by the pattern of divergence rests on what is today called the "competitive exclusion principle."

Simply put, this principle states that similar life forms are likely to compete more harshly with one another than dissimilar ones, if only because the former occupy nearly the same place in the economy of nature, thus putting them in a state of direct competition for nearly identical resources. To the question "which species are most similar to one another?," Darwin answers: those that share a common ancestor. That is why he argues, for instance, that two distinct varieties or species derived from a very recent common ancestor compete more vigorously with each other than two distinct forms not as closely related. He explains this in these words:

> But the struggle almost invariably will be most severe between the individuals of the same species, for they frequent the same districts, require the same food, and are exposed to the same dangers. In the case of varieties of the same species, the struggle will generally be almost equally severe ... As species of the same genus have usually, though by no means invariably, some similarity in habits and constitution, and always in structure, the struggle will generally be more severe between species of the same genus, when they come into competition with each other, than between species of distinct genera ... We can dimly see why the competition should be most severe between allied forms, which fill nearly the same place in the economy of nature. (Darwin 1859: 75–6)

Recent interpreters have tended to overlook the evolutionary implications of this aspect of Darwin's thinking, wrongly projecting our modern understanding onto it. Two such implications are considered below.

The Deactivation of the Evolutionary Drive

The first implication is brought out by paying closer attention to an idea found in the passage quoted above: "[T]he struggle will generally be more severe between species of the same genus ... than between species of distinct genera." Why should this be so? In more than one place in the *Origin of Species* (1859: 75–6, 110, 321). Darwin explicitly holds that forms less closely related face a lower level of competition as compared to more closely related ones. Evolutionary pressure, for Darwin, is gradually diminished as related forms become ever more different from one another. Indeed, if two closely related forms diverge away from each other under the impetus of the competitive exclusion principle, it follows that as they continue to evolve in distinct directions—each accumulating different sets of adaptations—they will compete less and less with each passing generation. On Darwin's own account, the impetus behind the *evolutionary drive* gradually weakens or diminishes with the ascent of the taxonomic scale.

A brief note of caution: the expression "evolutionary drive" employed throughout this book to describe the causal forces driving evolutionary change carries no mystic, vitalistic, or internal connotation, nor suggestion of inevitability, either for us or for Darwin. The expression merely refers to material causes involved in a given process that result in the selection of the best possible variations available in a particular place and time in the context of harsh competition.

For Darwin, then, when two forms begin their respective taxonomic climb away from a common ancestor—gradually becoming two distinct varieties, two distinct species, two distinct genera, two distinct families, even two distinct orders—the process should be accompanied by a weakening and ultimate deactivation of the underlying causes driving that very evolutionary process. So far, so good, for Darwin's evolutionary principle. But this raises a challenge for Darwin's theory: how can divergence be maintained under this scenario over the long course of geological time? Is it reasonable to assume, as Darwin does, that divergence could have been sustained over the entirety of evolutionary history? The obvious answer is "no"; on Darwin's *own logic* the evolutionary drive behind the process of divergence should eventually exhaust itself.

At the same time, no such evolutionary exhaustion can be discerned in Darwin's Diagram. Quite the contrary, one sees an overall divergent pattern equally in view at all taxonomic levels (Darwin, 1859: 116–26, 331–2, 412–13). Here, again, we should pay close attention to Darwin's way of putting forward arguments in these sections: while he seems to suggest that he has provided support for a given phenomenon, the careful reader discovers that Darwin himself offers good reasons not to follow him. In the present case, Darwin is clearly taking for granted that the competitive exclusion principle is operating in equal measure at all times and circumstances, applying even to forms quite removed from each other, genealogically speaking. Yet it is unclear why he should have held such a view. We suggest, therefore, that Darwin organizes evolutionary history around a giant "V" (divergence) not because of his commitment to the competitive exclusion principle, but rather because of his theoretical bias which assumes that all past and extinct forms *must* fall in-between currently existing ones, converging toward ancient common ancestors. The evolutionary show must go on, Darwin seems to say, with more and more divergence over time, come what may.

To be consistent, Darwin should have recognized that what is being gradually deactivated during taxonomic ascent is not necessarily the evolutionary drive per se but the adaptive cause for *two main related lineages*—each composed of a myriad of related forms—to continue to diverge from each other. Indeed, by the time these two main lineages independently reach the family or order level, for instance, the competitive exclusion principle may no longer be active enough to push them away (divergence). If selective pressure remains strong enough, there is no reason to believe divergence must be the only outcome; adaptation may also push them closer (convergence) or follow similar paths (parallel evolution). And if the selective pressure is too weak, these two main forms might just remain stable (stagnation), a theme discussed in our next chapter.

Again, Darwin's thinking about the past is permeated with, and structured by, the present time horizon. For him, the past is constructed entirely from his own historical standpoint; it is a fabricated illusion built exclusively from materials found only in the present. By overlooking the logical implications ensuing from his *own* explication of the competitive exclusion principle, Darwin committed the revealing inconsistency of assuming that this principle would be a constant force, equally effective at all taxonomic levels, precisely because he takes for granted that the conditions of the past are identical to those of today and will remain so in the future—an endless repetition of the "present"—projected backward and forward in time. Darwin here falls victim to his dismissal of the field of paleontology: his neontological biology, erected upon currently existing forms, effectively skirts any real confrontation with the past. Darwin's geometrical view of life's history, as seen in his foregrounding of the role of divergence, *negates* the historical dimension of evolution. As such, his view of time is based on the binary notion of extrapolation-interpolation: the future is a mere *extrapolation* of what is known today; the past is merely an *interpolation* of how things now appear. Darwin's view is essentially flat, lacking a genuine temporal dimension, as should be expected for someone embracing the static worldview.

The Elimination of Evolutionary Novelties

The second implication of Darwin's divergence-oriented view is its impact on the ability of the evolutionary process to find novelties or new solutions to adaptive challenges. Darwin is convinced that the future belongs to forms that diverge from others: "I attempted also to show that there is a constant tendency in the forms which are increasing in number and diverging in character, to supplant and exterminate the less divergent, the less improved, and preceding forms" (1859: 412). He argues that adaptation mostly occurs along evolutionary paths organized around opposing (diverging) forms, constraining evolution to finding *predetermined* solutions or forcing it to follow channelized evolutionary pathways.

To put it bluntly: why just divergence? Admittedly, Darwin's Diagram does reveal that he allows a *small degree* of convergence, but not to the point of undoing the overall divergent pattern of life. Indeed, some descendants of lineages A and I do come a bit closer to each other but none come even close to crisscrossing or exchanging taxonomic positions. Moreover, reticulate evolution is nowhere to be found in his diagram. Are not convergent and reticulate evolution also perfectly good survival strategies for life forms to adopt? These strategies were known to Darwin but largely dismissed by him. Contrary to what scholars of today often claim on his behalf, Darwin's theory does not present evolution as an open-ended process. Instead, he gives us a rigid and geometrical picture that cuts off many potential routes to evolutionary novelties, options now seen as essential to evolution.

We once again arrive at the same place, though by a different route: Darwin's belief that the taxonomic position of the main forms remains fixed throughout all of evolutionary history. Returning to the Diagram, we see that the relative position of lineages A, F, and I (including their descendants) remains the same since the beginning of life, with lineage F in the middle, A on the left side, and I on the right. How can Darwin justify this feature of evolution? Only by forcing his way through, theoretically speaking, and imposing a simplified vision upon evolution's complexities. Fully embracing convergent and reticulate evolution would have jeopardized his entire view, blurring the neat overall divergent pattern he envisioned.

Conclusion

Obstacles are piling up for those who would see Charles Darwin as a thoroughly modern evolutionist, perhaps even the very first one. The next chapter will allow us to lay out more clearly what, in our view, lies beneath Darwin's concealed commitment to a static worldview. For the moment, it should suffice to note that Darwin's overall divergence-centered account, whatever its explanatory merits (and there are some), comes at the cost of eviscerating both the temporal dimension of evolution and the perpetual rise of novelties that would accompany it, depriving evolution of two fundamental creative factors.

Annotated Bibliography

As in previous chapters, the two editions of the *Origin of Species* referred to in this chapter are Darwin, C. (1859), *On the Origin of Species*, London: John Murray; Darwin, C. (1872), *The Origin of Species*, 6th edn, London: John Murray.

As noted in the foregoing, Charles Darwin's Diagram is found in the middle of Chapter 4 in all six editions of the *Origin of Species*.

In the *Origin of Species* (1859: 63), Darwin offers a rather bleak, if not terrifying, image of nature presenting life forms as confronted with extreme adaptive challenges: "The face of Nature may be compared to a yielding surface, with ten thousand sharp wedges packed close together and driven inwards by incessant blows." See also 1859: 63–71, 80–7.

To understand Darwin's use of the "competitive exclusion principle," one must dig beneath the surface rhetoric and look closely at his arguments, which reveal why he himself overlooked the exhaustion of this evolutionary drive which he acknowledged. The reasoning comes in two parts. **First**, by organizing his "tree of life" around genealogically related forms, Darwin postulates that the more closely two forms are related, the more they will tend to diverge from each other, owing to the fact that they compete for nearly identical resources. This implies that the more they diverge, the less effective the selective impetus will be in pushing them away as they independently climb the taxonomic scale (from two varieties, to two species, to two genera, etc.). What is being gradually deactivated during taxonomic ascent, therefore, is not necessarily the evolutionary drive per se (which may or may not be sustained under new divergences associated with two specific types, say A1, A2, A3, A4 on the one hand, and B1, B2, B3, B4 on the other hand), but the adaptive cause for two related types to diverge from each other. By the time these two types independently reach the family or order level, for instance, the competitive exclusion principle is no longer active enough to *exclusively* push the two types away (divergence), initiating a sort of "random evolutionary walk" since adaptation may also push them closer (convergence), or follow similar paths (parallel evolution), or remain stable (stagnation). **Second**, as discussed elsewhere in this book, Darwin's agenda involves more than just the matter of neutralizing the competitive exclusion principle. He also embraced commitments that thwart the expression of the evolutionary drive altogether, effectively providing a list of reasons as to why evolution might be stopped. These include (1) an in-built system of opposed forces cancelling or balancing each other out, as seen in the progression/retrogression duality, the process of opening/closing evolutionary gates, in species nominalism versus species realism; (2) embryological laws grounded in fixed embryological types; (3) the significance of adaptive equilibrium; (4) the entrenchment of high taxonomic groups (classes and phyla) in their respective places in the economy of nature; (5) fixed taxonomic positions since the beginning of life. The list of such commitments is long and consistent, and all tend in the same direction, being expressions of a static worldview.

5

A Cyclical World in Equilibrium

Having noted Darwin's reluctance to embrace the dynamism of evolution in full, with respect to both its time dimension and evolutionary novelty, we now come to a set of questions at the heart of our investigation. Why did Darwin try to corral all past forms within the narrow confines of extant ones? Why would he deprive the evolutionary process of maneuvering room by putting it into a divergent straightjacket, to the point of thwarting the expression of evolutionary novelties? Why did he fail to realize that the competitive exclusion principle he himself postulated was ultimately self-defeating and would gradually weaken itself over time through taxonomic ascent? This chapter aims to answer these questions by reference to Darwin's deeply entrenched belief in the static worldview, a commitment on his part likely more unconscious than conscious. Indeed, while trying to work his way toward the evolutionary worldview, it seems that Darwin did not realize the extent to which he still clung to explanatory components typically associated with the static worldview. While a number of these components have already been discussed in our previous chapters, a new feature will be stressed here: the connection between the concept of motion and the static worldview we are here attributing to Darwin. Change is recognized in this latter view, but only in the limited sense of cyclical motion at the service of reestablishing the world's equilibrium. While this association seems rather strange to us now, it was not at all uncommon in the seventeenth and eighteenth centuries. As John C. Greene explains:

> Change was recognized as a real aspect of nature, but a superficial aspect. It contributed variety to nature's panorama, but it could not alter her fundamental structures. Change might take the form of decline from original perfection, of cyclical movement serving to maintain the status quo, or of random variation about a norm, but in no case could it produce real novelty.[1]

It is this conception of the world, and in particular the way change is thought to occur within it, that lies at the core of Darwin's view. Unpacking this concealed commitment accounts for a number of theoretical moves throughout the *Origin of Species* that would otherwise seem inexplicable, perhaps even strange. This strangeness cannot be laid entirely at Darwin's doorstep, however, since much of the ostensible oddity stems from our tendency to lose sight of the fact that evolutionary thought did not follow a single path in the history of ideas. Evolutionism and modern thinking about the universe in general came together as the result of the development of many disparate strands of thought, from scholars working in a range of fields. In the following sections we will focus on three of these strands, each which represent an attempt by thinkers of the past—Darwin's predecessors and contemporaries—to grapple with the disruption of traditional thinking about the universe wrought by the emerging idea of evolution.

The Scale of Complexity

Scholars have long been struck by the biological differences between currently living forms, some simple in organization (unicellular organisms), others much more sophisticated (multicellular organisms with specialized parts). When the idea of biological evolution began to gain some traction in the late eighteenth and early nineteenth centuries, it occurred to a number of scholars that differences in levels of organization might be explained by a process of complexification over the long run of geological time. The French biologist Jean-Baptiste Lamarck, for instance, in his *Zoological Philosophy* (1809), suggested that complex forms such as humans had reached a high level of complexity because they had a head start, beginning their evolutionary ascent much earlier than simpler creatures who had much less time to climb the scale. For Lamarck, living forms were not bound together in common ancestry; instead, they each had to make their own way up the scale of complexity. A somewhat similar system was presented by the Scottish scholar Robert Chambers in his *Vestiges of the Natural History of Creation* (1844).

A Discontinuous History of Life

Another line of thought relied extensively on the newly emerging field of paleontology in proposing a view of life organized around distinct and separate episodes, groups of fauna and flora that once dominated the Earth only to disappear from the fossil record and be replaced by new sets of forms. Under this interpretation, the history of life appears as a rising and falling succession of specific floras and faunas, separated in geological time by profound discontinuities and replacement events, with ancient forms finding few counterparts among extant ones. As already alluded to, this approach was fostered by Georges Cuvier and Alexandre Brongniart's *Description géologique des environs de Paris* (1825) and William Buckland's *Geology and Mineralogy Considered with Reference to Natural Theology* (1836).

A Steady-State World

Charles Darwin's approach to life's history differed in several important ways from the two schools of thought just described. First, despite its title, the *Origin of Species* is not focused on the rise of simple life forms. Aside from a handful of abstract and vague remarks (Darwin, 1859: 484, 488, 490) to the effect that life may have arisen from a limited number of truly primitive progenitors, Darwin's biology works backward from fully constituted forms now alive. As Robert M. Young concludes: "[Darwin's] doctrine, like those of Hutton and Lyell before him, was not concerned with the origin of earth or of life."[2] Second, however much Darwin may insist that life is united under the principle of "descent," the fact remains that his biology is actively constructed around separate and distinct classes (such as fish, reptiles, and mammals) and phyla (e.g., vertebrates, mollusks, and sponges). There is no unified view of life in Darwin; only a segmented tree of life, that is, a collection of separate trees of life each representing a class (see Figure 13). Third, Darwin practiced what may be called an "epistemology of assimilation": no matter what he encountered in the annals of life, he made sure that no ruptures or discontinuities would be recognized by presenting us with a past characterized by a unified, single, and continuous network of forms, but only within each distinct class or phylum, implying that a tight connection between the past and the present exists within such separated entities (Darwin, 1859: 489).

As Greene explains in the quotation above, proponents of the static worldview did not hold that the world was literally motionless; only that motion was to be explained away rather than taken at face value. Isaac Newton's view of the universe was the most influential incarnation of this idea in the eighteenth century: quite a bit of motion is acknowledged to exist in our solar system in the form of the planets revolving around the Sun. At the same time, the solar system is perfectly stable or static while simultaneously incorporating motion within it. For Newton, it was pointless to look to the past in search of potentially different states of the world. What can be observed today is what existed in the past and what will exist in the future, unless decided otherwise by the Creator. To put it another way: there is no more to the history of the world than can be observed today; the history of life, one might say, is nothing more than a succession of present times. For Newton as for others to be discussed shortly, the world is characterized by its *steady state*.

Considering that the notion of a "steady-state" world is central to this chapter, it will be useful to provide a brief definition. It signifies a stable system that remains essentially the same over time, wherein any change in one direction is perpetually counterbalanced by a change in another direction, thus maintaining the system in equilibrium. It is obvious why the notion of "steady-state" is congenial to a "static worldview" in which cyclical motion is allowed. Ultimately, the notion of "uniformitarianism" discussed earlier also fuses with the latter two notions, forming a mutually reinforcing conceptual bundle. Processes and entities are seen as similar and uniform across temporal horizons, with the past, the present, the future being identical in fundamental ways.

How did Darwin manage to carry this kind of thinking with him into the nineteenth century at the precise moment when the idea of evolution was steadily making its way in the minds of an increasing number of scholars? In a nutshell, Darwin ensured that the stability of the biological system would be kept intact by the immutability of the classes and phyla. These higher taxonomic entities have been in existence, Darwin holds, for as far back as one can probe into geological time. Lest Darwin's system break under the intellectual pressure of the rapidly developing doctrine of evolutionism, however, he had to bend it by allowing for some degree of evolutionary motion (change). Darwin ensured it did not collapse entirely by conceding a degree of motion within certain limits, that is, among lower taxonomic entities (species, genera, families, and orders) belonging to the same class or phylum. He acknowledged that these might become extinct over time and be replaced by other such low-level taxonomic entities. Evolutionary motion is thus introduced into the system, in a circumscribed fashion, by allowing for the rise and fall of lower taxonomic entities—a sort of cyclical motion—while the system as a whole remains static, as seen in stable and permanent classes and phyla.

Like Newton, Darwin assumed that the present time represented the ultimate time horizon for understanding the world. Whereas Newton lived at a time when most people would not have even contemplated the possibility of a substantively changing world, this began to be seriously questioned in the nineteenth century. At some level Darwin was committed to salvaging the Newtonian conception of time and change, then being assailed on various fronts. He did so by shoehorning the past into the present. As the ground beneath his feet began to crumble—revealing a much richer and variegated history of life documented by paleontology—Darwin found ways to assimilate these findings to extant forms. For Darwin, the past could easily be dismissed, since it is in principle less informative than the present and provides only sketchy and disconnected bits of data. Darwin preferred instead to *construct*, rather than *discover*, "ancestors" by projecting features selected from among extant forms backward in time. As discussed earlier, these "ancestors" are chimera of a sort, built out of bits and pieces of actual forms, and are ultimately shadows or pale reflections of extant forms. This is yet another manifestation of Darwin's cyclical view of the world, one which recycles identical variations by projecting them backward in time. By adopting this approach, he was able to remain true to the spirit of a steady-state view. His commitments in this regard are also reflected in the fact that only two chapters out of thirteen of the *Origin of Species* (1859) are devoted to the history of life, while all others focus on issues concerning currently existing forms. Insofar as he discusses the history of life at all, it is to downplay the potential contributions of paleontology to the study of evolution. His dismissive attitude toward fossil evidence speaks volumes about the way he conceives of the relationship of past and present and puts him directly at odds with modern evolutionary theorists (and, as we have seen, even with some of his contemporaries).

Geological and Biological Worlds in Equilibrium

In order to make Darwin's fundamental commitment more intelligible to us, it will be useful to trace out the tradition of thought to which he belonged by highlighting some of its key representatives. Newton's static celestial mechanics in perpetual motion, discussed above, would ultimately reappear in the fields of geology and biology. One can see its influence in the study of geology, for instance, in James Hutton's *Theory of the Earth* (1795), which envisions an Earth whose parts are both wearing out and renewing themselves in equal measure, in a never-ending cycle. Roughly speaking, Hutton sees things working along these lines: (1) rivers and ocean waves erode the surface of the Earth; (2) the eroded debris is deposited in stratified geological layers, with their accumulated weight generating heat in the lower strata; (3) the heating of lower strata continues until new land surfaces are forced upward. In a telling phrase, Hutton summed up his view of Earth's history with the famous motto "we find no vestige of a beginning—no prospect of an end."[3] A key element of Newton's celestial mechanics was thus retained by Hutton's geological system in the form of the idea of a prevailing equilibrium.

This same idea also penetrated and persisted in the world of biology. The Swedish scholar Carl von Linnaeus, the creator of our present-day system of taxonomy, conceived population changes on Earth as a slowly expanding process unfolding over historical time. Taking the Great Flood as a starting point in his *Oratio de Telluris habitabilis incremento* (1744), he imagined an original island populated by a single pair of each sexually reproducing species. Eventually, newly habitable territories opened up to them as the water level receded. At the same time, a state of equilibrium is maintained throughout this process of colonization. Indeed, nothing genuinely new is introduced, as the same relationships between species are maintained and regulated in advance such that encroaching species would not drive others to extinction. In *Oeconomia naturae* (1749) and *Politia naturae* (1760), Linnaeus describes a "world machine" in motion, but a circular motion with no beginning or end, for cyclicity permeates nature and is its guiding principle: the birth and death of life; the propagation-preservation-destruction cycle in the three interrelated kingdoms; the perpetual circulation of water (evaporation and rain); the alternation of seasons and of day and night.

It is worth noting that in inheriting the flood myth from the Book of Genesis, Linnaeus took on board not only its account of events, but also a way of understanding the world; in this case, an approach to the world that was deeply imbued with biblical theology. As accustomed as we now are to thinking of "revolutions"—through such events as the Industrial Revolution, the American Revolution, and Computer Revolution—the greatly oversimplified notion of wholesale shifts in thought may be part of our own mythology, linked in some ways to the myth of the "founding figure" (an analogue of the "Great Man Theory" in history) discussed earlier. Our current understanding of the puzzle of nature is rather more complete thanks to the work of Newton, Linnaeus, and Darwin, but these scholars' ideas did not arise out of the void in a single flash of insight or inspiration but took shape slowly; bits and pieces of novelty and innovation were introduced and made to coalesce with other, older, unexamined bits. This is characteristic of the development, or if you like, the "evolution," of thought, and in the case of each of these three men, we should not underestimate the degree to which a theocentric account of the universe shaped their thinking, directly or indirectly. The static understanding of nature is one such inherited presupposition. While Darwin is currently seen as a kind of hero or champion by those imbued with a great admiration for science, or those aiming to use it as a replacement for theology, it is a significant distortion to overlook how deeply indebted he was to a static worldview, a notion itself not without theological connotations. As Robert M. Young observes, "theology moved toward the identification of God with the uniformity of nature,"[4] meaning that the laws of nature presumably created by God became more and more seen as being associated with, or incarnated by, uniformitarian ideas as the nineteenth century progressed. This impugns neither Darwin nor his contribution: it simply registers the fact that, to again paraphrase Newton, even geniuses stand on the shoulders of giants. This reflection, we hope, will help to inspire a measure of healthy skepticism regarding the supposition that Darwin was the first *modern* evolutionist, casting him instead as a transitional figure between ancient and modern thinkers.

Returning to Linnaeus: we can see, on his view, how the "order" currently observable in the world is something to be explained, over and above the world as we find it. The world's order is preserved by a series of checks and balances: for example, herbivorous animals, by eating plants, are preventing the latter from overpopulating; carnivorous animals, in turn, keep herbivore populations in check. Linnaeus saw nature as always full to capacity, with any open spaces immediately occupied by already-existing species, with some species or other always standing ready to fill any vacuum. In this "war of all against all"—an expression used by Linnaeus—each living form maintains itself at the cost of confronting another. This image of nature as a world perpetually at war was shared among many thinkers of the day. It is thus unsurprising at one level that the work of Linnaeus written in the eighteenth century should be so strangely reminiscent of some pages found in the nineteenth-century Darwin's *Origin of Species* (1859: 71–9).

Facing the Reality of Extinction

Everything is well ordered in the fully constituted world of Newton, Hutton, and Linnaeus. After all, the static worldview implied that the motion observed in astronomy, geology, and biology was perfectly cyclical—a mere manifestation of a world in equilibrium with the past and future being identical to the present.

However, this does not take into account the rise of a major challenge first appearing in the late eighteenth century and gradually gaining force in the early nineteenth century, up until the publication of the *Origin of Species*: the growing evidence that many life forms eventually become extinct. As the field of paleontology gathered momentum during that period, it became more and more difficult to ignore the fact that some forms now inscribed into the annals of life had no apparent living counterparts. Growing recognition of this fact posed a clear threat to a biology built on the static worldview. How could this be possible if the past and future are identical to the present? How could this be possible if Earth and the rest of universe had no history beyond what can be observed today?

At first, it seemed that the static worldview could be saved by simply denying the reality of extinction altogether. Surely, the argument went, these previously unknown forms excavated from the past *must* have currently living representatives that we have yet to discover, *somewhere*, in some remote and neglected corners of our planet. John Ray, Carl von Linnaeus, James Hutton, and even Thomas Jefferson all took this approach. This view has quite a bit more going for it when viewed from the perspective of the time: the "known" world, even in geographic terms, was much smaller than we know it to be today. With so much undiscovered territory, and new discoveries streaming in from all over, the world may have seemed to many like an open book. Be that as it may, this attempt at a solution did not fare well; not only would a thorough search of Earth's surface fail to turn up such living forms, but paleontology would also continue to churn up even more, and ever more strange and unique, forms. This might have spelled the end of the road for the static, steady-state worldview, but a way out remained for those hesitant to let it go. One could argue that while the past and the present may not be identical in *every detail*, they remained essentially the same with respect to their *main features*. This strategic retreat insured that the present would remain the proper horizon for understanding the whole of Earth's history. In describing this theory-saving tactic we do not wish to imply dishonesty; merely that these scholars were loath to give up deeply held beliefs or were simply unable to see things any other way.

It was thus left to the heirs of Newton, Hutton, and Linnaeus to contend with the threat brought by the advance of the burgeoning science of paleontology. In the strategic retreat to recast the static worldview, Charles Lyell's *Principles of Geology* (1830–1833) adopted a sort of intermediate position between Hutton and Linnaeus in the eighteenth century and Darwin in the nineteenth. Lyell embraced a steady-state geological system composed of counterbalancing forces: so-called *aqueous factors* (glaciers, rivers, and streams) erode the surface of the Earth, while *igneous ones* (volcanic activity) replenish it. While on this view eroded and restored features are not replaced by perfectly identical ones—exact replicas of lakes, seas, continents—this remodeling of Earth's surface maintains a state of equilibrium when viewed both *as a whole and over time*. Lyell's application of a steady-state system to biology is especially interesting in this regard. Acknowledging that the reality of extinction could no longer be denied, he adopted what might be called a "minimalist" position: while affirming the extinction of *individual species* faced with ever-changing environmental conditions, he simultaneously refused to accept that there was once a time when *main biological types* seen today—mollusks, fish, mammals—did not exist. In other words, he accepted the extinction of forms at very low taxonomic levels but not at higher ones. Lyell's uniformitarian view, therefore, was left with no way to account for the rise of real novelties in the history of life: the main features of life have been in place for as far as one can probe the geological past, he claimed. Lyell's system, however, came with the significant disadvantage of implying a gradual depletion of biomass over time—the loss of individual species—for which no explanation of replenishment was provided.

It is at this juncture that Charles Darwin enters the picture with a partial solution to the rigidity imposed by the steady-state system: like Lyell, he allowed for the disappearance of individual species but moved things up a notch by recognizing that genera, families, and orders also went bust. Darwin was thus able to introduce a little more explanatory flexibility into both Linnaeus's and Lyell's systems by allowing for an added measure of motion. This recasting of the steady-state system in biology involved three distinct moves:

1. Linnaeus postulated that each particular species is uniquely adapted to fit into a single physical place or station in the economy of nature. Darwin had to dissolve this one-to-one correspondence by postulating, instead, that species are more adaptable than this and not necessarily confined to a single physical space.
2. Linnaeus argued that the world system was founded on preestablished relationships between species, without which the balance of nature would be upset by some species driving others to extinction. Darwin turned this reasoning on its head: equilibrium exists in the world not because of preestablished relationships, but rather as the result of the various

confrontations between life forms. Equilibrium is thus the outcome of such confrontations between warring species: dominating forms eventually imposing themselves upon dominated ones.

Lyell paved the way for Darwin on this matter by arguing that species in the geological past were continually pushed around by perpetually changing environmental conditions, leading to occasional extinctions of particular species. What was a gradually self-depleting world for Lyell, however, was in Darwin's hands a creative opportunity: the removal of extinct forms opened up new possibilities for remaining ones.

3. Linnaeus organized the living world around unrelated forms whose role consisted in preventing one from dominating the other. For his part, Darwin organized the living world around forms bound together in common ancestry (genealogy). This is illustrated in his Diagram (see Figure 9). Ever more divergent forms expand over time, usually driving less divergent forms to extinction.

Darwin's world system succeeded in creating more motion and maneuvering room for life. Yet this world system was still far from being truly open in the sense we now understand. Indeed, Darwin made sure that it would be locked up and closed in on itself by allowing a limited degree of motion at lower taxonomic levels, and little or none at higher ones. As one would expect from someone pushing the conceptual envelope of the static worldview, the present time horizon always remains the privileged vantage point from which to look at the past. For Darwin, as we have argued, the present always reveals more than the past, the latter being merely a diminished version of the former. This is reflected in several explanatory devices enshrined in his system. For present purposes it will suffice to repeat the most important of these, which have been discussed in the previous chapters:

1. All the biological diversity of extinct forms is believed to fall in-between that of extant forms.
2. Common ancestors of currently existing forms are imagined by projecting variations seen among extant forms backward in time.
3. The taxonomic arrangement currently observed among living forms is said to have maintained itself throughout the entire history of life.
4. The history of life is segmented by being organized around separate and independent classes and phyla.

While species, genera, families, and orders within each class or phylum are free enough to rise and decline (extinction)—replacing each other over time—this whole process occurs within the confines of separate and permanent classes and phyla profoundly entrenched in nature since life's beginnings.

Darwin was not the only proponent of this updated version of the steady-state worldview at the time of the publication of the *Origin of Species* in 1859. Two of his countrymen had arrived at similar conclusions. For instance, in his 1862 address, Thomas Henry Huxley said:

> The positive change in passing from the recent to the ancient animal world is greater, but still singularly small. *No fossil animal is so distinct from those now living as to require to be arranged even in a separate class* from those, which contain existing forms. It is only *when we come to the orders*, which may be roughly estimated at about a hundred and thirty, that we meet with fossil animals so distinct from those now living as to require orders for themselves; and these do not amount, on the most liberal estimate, to more than about ten per cent of the whole.[5] [our emphasis]

Similarly, Richard Owen wrote in 1860:

> *Every known fossil belongs to some one or other of the existing classes*, and that the organic remains of the most ancient fossiliferous strata *do not indicate or suggest that any earlier and different class of being remains to be discovered*, or has been irretrievably lost in the universal metamorphism of the oldest rocks.[6] [our emphasis]

These remarks are rather similar to what Charles Darwin had to say in the *Origin of Species* (1859: 129 30):

> From the first growth of the tree, many a limb and branch has decayed and dropped off; and these lost branches of various sizes may represent those whole *orders, families*, and *genera* which have now no living representatives, and which are known to us only from having been found in a fossil state. [our emphasis]

For Darwin, then, no individual *class or phylum* ever goes extinct; only orders, families, genera, and species do. While a number of his contemporaries were busy organizing the history of life either around a scale of complexity or in terms of discontinuous episodes, Lyell, Darwin, Owen, and Huxley all worked their way *backward* in geological time by projecting fully constituted classes into the past. Seen in this light, one is suddenly struck by the numerous instances by which phenomena are explained in terms of the dual notion of "equilibrium-and-cycle." In a sense, this is to be expected: if the history of life is not an open-ended process, motion in a closed world must somehow be either accommodated or dissipated.

Equilibrium and Stagnation

Speaking of competing forms living under constant pressure from each other, Darwin writes: "Battle within battle must ever be recurring with varying success; and yet in the long-run the forces are so nicely balanced, that the face of nature remains uniform for long periods of time" (1859: 73). Darwin's world is, at bottom, Linnaeus's world, one that is filled to capacity with encroaching forms that lock each other up in the economy of nature. As previously noted, a somewhat strange feature of Darwin's Diagram now finds its full rationale: nine out of eleven lineages are presented as moving along in a state of perfect evolutionary stagnation, before eight of them eventually go extinct after more or less extended geological periods (see lineages B, C, D, E, F, G, H, K, and L in Figure 9). The notion promoted by Darwin here could well be called "adaptive equilibrium," which sees closely related forms as profoundly entrenched in their respective places, confronting each other and standing their ground in mutual stability:

> But we often take, I think, an erroneous view of the probability of closely allied species invading each other's territory, when put into free intercommunication. Undoubtedly if one species has any advantage whatever over another, it will in a very brief time wholly or in part supplant it; but if both are equally well fitted for their own places in nature, both probably will hold their own places and keep separate for almost any length of time. (Darwin, 1859: 401)

Furthermore, having organized life forms around common ancestry, Darwin was confronted with a difficulty. We have seen that a cardinal notion of his theory was that competition between very closely related forms was harsher than that between more distant forms, if only because the numerous similarities of the former make them more prone to exploiting the same resources in nature. Darwin relied heavily on the "competitive exclusion principle" to explain why such related forms would diverge away from each other. Strangely, however, Darwin himself gave reasons as to why very closely related forms could not easily be disentangled from the common reproductive network to which they belong. If close races, varieties, and species are connected by their ability to reproduce among themselves, how can they be induced to break free from this genealogical network in the first place?

For Darwin, the forces resisting change and, ultimately, segregation of species (divergence) are numerous. Before enumerating the most important ones, the reader should again be made aware that at points, Darwin tries to have it both ways with respect to one of the most important tensions or contradictions to be found in the *Origin of Species*. To put this bluntly, Darwin holds, on the one hand, that he has convincingly demonstrated that life is significantly organized around the principle of divergence (with life forms fanning out from common ancestors), whereas, on the other hand, the reader is offered in the *Origin* a long list of reasons as to why life forms cannot diverge away from each other. Darwin's reasons are the following (Darwin, 1859):

1. Life is described in terms of non-monolithic taxonomic entities called "strains" and "sub-breeds"; that is, entities inherently characterized by their ability to bind through reproduction (pp. 31, 32, 96, 267).
2. Life naturally favors interbreeding over close inbreeding, meaning that in mating, forms tend to seek out forms not too closely related (pp. 70, 71, 96, 104–6, 248–50, 253).
3. Life is characterized by various levels of fertility below the family level, that is, at the taxonomic levels of races, varieties, species, and genera (pp. 22, 23, 248, 253, 255–7, 261, 267–72).
4. Breeders are able to create forms separated at the generic level yet still fully fertile with respect to one another (pp. 22, 23, 26, 445).
5. The recognition that the derivation from more than one ancestral line (polyphyletism) constitutes a common genealogical feature in domesticated forms (pp. 17, 28, 29, 40).
6. Darwin's support for a heredity that operates by blending, that is, by mixing together features of both parents in generating their children, thus achieving intermediate states (p. 108).
7. Life resisting change for several generations (p. 7).
8. Descendants having the tendency to revert to their ancestral conditions (pp. 14, 15, 25, 26, 152–4, 159).

Confronted with this list of commitments that run directly counter to the segregation of related forms (divergence), Darwin scholars have long recognized his failure to provide a convincing explanation of the division or speciation process (see also Darwin 1859: 102–8). Here, Darwin fell victim to his own approach to envisioning the past, projecting present features backward in geological time. The *Origin of Species* devotes significant attention to studying related extant forms at low taxonomic levels (individuals, varieties, species, and genera). Looking through that lens, Darwin was struck by how life forms were entangled in complex web of reproductive networks. This view was then projected into the past, creating the impression of an uninterrupted network of life forms organized in a tightly knit fabric. For Darwin, the fabric of life is seamless, a notion perfectly summed up in his oft-repeated motto in the *Origin* (1859: 194, 206, 210, 243): *Natura non facit saltum* ("nature makes no jumps").

If nature is indeed a seamless fabric devoid of biological gaps and discontinuities, one should expect to see a fair number of intermediate or transitional forms lying between any two given forms. Darwin thus subscribed to what we will now call the "principle of gradation." This principle is intimately fused with another principle at the core of his theory, the aforementioned "principle of divergence" outlined in previous chapters. Once again, this former principle plays the same explanatory role as the latter: linking the past to the present. Indeed, while the principle of divergence *brackets* past forms—they all fall in-between extant forms—the principle of gradation tightly *connects* past forms to present ones in a contiguous fashion. We now reach one logical conclusion of our investigation concerning Darwin's commitment to the notion of "equilibrium": related life forms encroaching each other are so tightly pressed together in Darwin's vision that no wiggle room exists for them to break free from one another (divergence), thus thwarting the rise of evolutionary novelties. The idea of equilibrium, in Darwin, is closely linked to the notion of stagnation or prolonged stability.

Equilibrium and Counterweights

The idea of equilibrium also finds another expression in the *Origin of Species*: for each possible evolutionary step forward, a counterbalance or opposing force is imagined. Let us review how Darwin applied this idea. First, for decades, Darwin scholars have debated whether or not Darwin truly supported the idea of biological progress, and whether or not this notion is compatible with his overall theory. Darwin was anything but clear on this issue (hence the debate) and instead offers the reader a mixed bag of inconsistent, sometimes even contradictory, arguments (1859: 314, 336–7, 345, 490). These can be classified under three main headings:

1. Because "biological progress" is an ill-defined notion, it is difficult to capture. In addition, it is impossible to test whether or not more recent forms are more progressive than more ancient ones, among other things. Anthropocentrism also introduces a possible bias into any specification of "progress" in this context.

2. The history of life reveals the production of higher forms in the scale of organization under the process of differentiation and specialization driven by natural selection.
3. Since there is "no fixed law of development," some lineages will change somewhat and others not at all; change may also be expressed through the simplification of forms over time, while the stability of simple forms may remain so indefinitely.

Darwin was of course well aware that in the history of life, progress of some kind had to be recognized. After all, without it, life would have remained stuck at the stage of unicellular organisms. At the same time, his conception of life is not organized around an ascending scale of complexity as in some contemporaneous views (as noted earlier). His approach consisted instead in projecting already fully constituted and complex classes backward in time. So, while Darwin did subscribe to a notion of biological progress, his profound commitment to the static worldview limited his efforts on its behalf. In the end, this view owed more to rhetoric than demonstration.

Darwin's view is in a perpetual state of tension between life moving forward and life at a standstill, frozen in time. This tension becomes apparent under a duality of opposing notions: "progression" versus "retrogression." For instance, one reads in the last edition of the *Origin of Species* (1872: 98):

> [B]earing in mind that all organic beings are striving to increase at a high ratio and to seize on every unoccupied or less well occupied place in the economy of nature, that it is quite possible for natural selection gradually to fit a being to a situation in which several organs would be superfluous or useless: in such cases there would be retrogression in the scale of organization ... On our theory the continued existence of lowly organisms offers no difficulty; for natural selection ... does not necessarily include progressive development—it only takes advantage of such variations as arise and are beneficial to each creature under its complex relations of life.

The progression/retrogression duality is essentially a reformulation of the concept of the *balance of nature* we have seen operating in the thought of Linnaeus in the eighteenth century, modified to suit a nineteenth-century context requiring a little more evolutionary movement: whereas some forms may become more complex, others may go through a simplification, as if to compensate. The ultimate underpinning of this conceptualization can be seen in Darwin's evolutionary dynamics, which appeal to a geometrical relationship between life forms: population "Y" will expand in space and time only if population "Z" moves out of its way (extinction). In a world filled to capacity, expansion of one form can only be achieved at the expense of another. According to Darwin, the number of forms during the history of life has essentially remained the same:

> Thus the appearance of new forms and the disappearance of old forms ... are bound together. In certain flourishing groups, the number of new specific forms which have been produced within a given time is probably greater than that of the old forms which have been exterminated; but we know that the number of species has not gone on indefinitely increasing, at least during the later geological periods, so that looking to later times we may believe that the production of new forms has caused the extinction of about the same number of old forms. (Darwin 1859: 320)

This rigid mechanical equilibrium of nature seems anything but congenial to the evolutionary flexibility required by the idea of "evolutionary contingency" so often associated with Darwin's name. As also seen in the progression/retrogression duality, the evolutionary gates open and close in a *remarkably symmetrical inverse relationship*, with some entering only as others exit.

Another issue long debated among Darwin scholars gains a fresh perspective when seen in light of Darwin's heavy orientation toward equilibrium: Darwin's conception of what it is to be a "species." Known as the "species question," the issue centers on whether Darwin held a "realist" view of species or a "nominalist" view. A nominalist stance takes entities to be so fleeting as to possess no fixed essence, as in the case of always-evolving species changing in space and time. In contradistinction, the realist position sees entities as characterized by a fixed essence, explaining why this doctrine is less suited to dealing with evolving species. Interestingly, Darwin wavered between the two doctrines. Indeed, Darwin (1859: 51) often speaks as if species were truly fleeting and transitional entities moving imperceptibly from one state to another, thus embracing nominalism. At other times, however, Darwin (1859: 177–8) describes species as if they were well-defined entities occupying stable places in the economy of nature, suggesting that Darwin inclined toward realism.

Here again, the point is not to situate Darwin in one of these two camps, but to note, in light of what we have argued throughout this book, that his vacillation between the two is entirely unsurprising given the balancing act Darwin was needing to pull off between the static worldview and the evolutionary worldview. It is yet a further manifestation of the profound tension in the *Origin of Species* discussed earlier, that is, the tension that drives Darwin to alternate between a world characterized by profound and ubiquitous change (nominalism) and a world devoid of it (realism). It should also be noted that, between the two poles, Darwin seems to incline more toward species realism than species nominalism, inasmuch as he sees higher taxonomic entities like classes and phyla as stable, permanent, and deeply entrenched in nature.

One final manifestation of the idea of *equilibrium* in Darwin will require our attention here. Under the competitive exclusion principle, Darwin organizes evolutionary dynamics around yet another *symmetrical but inverse relationship*. He postulates that evolutionary divergence is the outcome of closely related forms competing for similar resources; moving away from each other is the adaptive response organisms employ to reduce the level of competition. As we have noted above, Darwin's reasoning implies that while closely related varieties and species compete harshly with one another, competition is reduced at the higher taxonomic levels of genera, families, and orders. The *Origin of Species* (1859: 75–6, 110, 321) is quite explicit on this point. The underlying logic is as follows: *the evolutionary drive pushing two related forms (or two groups of forms) to diverge from each other will gradually exhaust itself in direct and reverse proportion to their taxonomic ascent.* In other words, the more two related types become separated through divergence—by becoming two related species, two related genera, or two related families—the weaker the effect of the competitive exclusion principle. Again, this is perfectly consistent with Darwin's *realist* side, which holds that high taxonomic entities like classes and phyla are stable, permanent, and deeply entrenched in the book of nature.

Conclusion

Darwin cannot rightly be thought of as the first modern evolutionist. He is better seen as a transitional figure, one of the last defenders of the static and steady-state worldview. His originality lies in how he modified this system to make it fit with evolutionary biology by creating more room for evolutionary motion through the rise and fall of varieties, species, genera, families, and orders, albeit without extending the privilege to classes and phyla. Something deep in Darwin's thinking opposed and resisted the notion of open evolution. In addition to the examples presented in the previous chapters of this book, we have just seen in this chapter two additional key aspects in the *Origin of Species*: first, in the recognition of the importance of phenomena like evolutionary stagnation and adaptive equilibrium; second, in the built-in system of opposed forces cancelling or balancing each other out, as seen in the progression/retrogression duality, in the process of opening/closing evolutionary gates, in species nominalism versus species realism, and in the exhaustion of the evolutionary drive in reverse proportion to the taxonomic ascent (the deactivation of the competitive exclusion principle).

Annotated Bibliography

The ideas of *cycle* and *equilibrium* are fundamental to this chapter. Several critical scholars have caught a glimpse of this aspect of Darwin's work in the *Origin of Species*, such as Beer, G. (2009), *Darwin's Plots*, 3rd edn, Cambridge: Cambridge University Press, 116–18. Working toward different ends, and under the pervasive influence of what we have termed the *standard view*, the full implications of these notions for reinterpreting the work of Darwin are not fully drawn out in these works. The present chapter is an attempt to take a further step toward that critical reinterpretation.

As elsewhere, the two editions of the *Origin of Species* referred to in this chapter are Darwin, C. (1859), *On the Origin of Species*, London: John Murray; Darwin, C. (1872), *The Origin of Species*, 6th edn, London: John Murray.

For a historical analysis on how motion can be accommodated within a static worldview, consult Greene, J. C. (1957), "Objectives and Methods in Intellectual History," *The Mississippi Valley Historical Review*, 44: 58–74. This text is reproduced in Greene, J. C. (1981), *Science, Ideology, and World View: Essays in the History of Evolutionary Ideas*, Berkeley: University of California Press, 9–29.

For a useful overview of the complex rise of the idea of evolution in the eighteenth and nineteenth centuries, we again refer the reader to Bowler, P. J. (2003), *Evolution: The History of an Idea*, 3rd edn, Berkeley: University of California Press.

The development of the field of paleontology in the nineteenth century is too complex to be profitably tackled here. Particularly helpful resources include Bowler, P. J. (1976), *Fossils and Progress: Paleontology and the Idea of Progressive Evolution in the Nineteenth Century*, New York: Science History Publications; Rupke, N. (1983), *The Great Chain of History: William Buckland and the English School of Geology (1814–1849)*, Oxford: Oxford University Press; Laurent, G. (1987), *Paléontologie et évolution en France, 1800–1860: Une Histoire des idées de Cuvier et Lamarck à Darwin*, Paris: Editions du Comité des Travaux historiques et scientifiques; Rudwick, M. J. S. (2008), *Worlds before Adam: The Reconstruction of Geohistory in the Age of Reform*, Chicago: University of Chicago Press.

James Hutton's famous adage "we find no vestige of a beginning—no prospect of an end" is found in his 1788 paper "Theory of the Earth," *Transactions of the Royal Society of Edinburgh*, 1: 209-305. This was also a title of a later book published in 1795.

It should be noted that Linnaeus used the expression "war of all against all" no less than three times in his *Politia naturae* (1760). This expression often exclusively associated with Darwin's view of life was, in fact, not uncommon in the eighteenth century.

Charles Darwin is often believed by laypersons to be a straightforward materialist and atheist, but this does not accurately reflect the complexity of his thinking. While he may have been a religious person when young, his sentiments on the question certainly evolved toward agnosticism as he grew older. His lost or diminished faith in Christian dogma, however, did not necessarily entirely erase a sense of the sublime and of moral virtues. As Nietzsche and others have reminded us, one can abandon religion and still retain metaphysical beliefs rooted in a theocentric conception of the world. Consult Young, R. M. (1985), *Darwin's Metaphor: Nature's Place in Victorian Culture*, Cambridge: Cambridge University Press, 126-63; Brooke, J. H. (1991), *Science and Religion*, Cambridge: Cambridge University Press; Sloan, P. R. (2001), "The Sense of Sublimity: Darwin on Nature and Divinity," *Osiris*, 16: 251-69; Richards, R. J. (2009), "Darwin's Theory of Natural Selection and Its Moral Purpose," in M. Ruse and R. J. Richards (eds), *The Cambridge Companion to the* 'Origin of Species', 47-66, Cambridge: Cambridge University Press; Brooke, J. H. (2009), "Laws Impressed on Matter by the Creator? The Origin and the Question of Religion," in M. Ruse and R. J. Richards (eds), *The Cambridge Companion to the* 'Origin of Species', 256-74, Cambridge: Cambridge University Press; Brooke, J. H. (2009), "Darwin and Victorian Christianity," in J. Hodge and G. Radick (eds), *The Cambridge Companion to Darwin*, 197-218, Cambridge: Cambridge University Press; Levine, G. (2011), *Darwin the Writer*, Oxford: Oxford University Press, 6-8.

Still a valuable source of information about the rise of the fields of geology and paleontology and the reality of extinction is Greene, J. C. (1959), *The Death of Adam: Evolution and Its Impact on Western Thought*, Iowa: Iowa State University.

For an excellent overview of Charles Lyell's *Principles of Geology*, consult Rudwick, M. J. S. (1990), "Introduction to the Re-edition of Lyell's *Principles of Geology*," Chicago: University of Chicago Press, vii-lviii. In the 1820s, 1830s, and 1840s, Lyell was an unambiguous anti-evolutionist. By the 1860s, he began to open up to the idea of evolution. See Bartholomew, M. (1973), "Lyell and Evolution: An Account of Lyell's Response to the Prospect of an Evolutionary

Ancestry for Man," *British Journal for the History of Science*, 6: 261–303; Gould, S. J. (1987), *Time's Arrow, Time's Cycle: Myth and Metaphor in the Discovery of Geological Time*, Cambridge: Harvard University Press, Chapter 4.

It is well known that Charles Darwin was profoundly influenced by Charles Lyell. The classic exposition of this thesis is found in Hodge, M. J. S. (1982), "Darwin and the Laws of the Animate Part of the Terrestrial System (1835–1837): On the Lyellian Origins of His Zoonomical Explanatory Program," *Studies in History of Biology*, 6: 1–106. Whereas Hodge holds that Darwin's primary inheritance from Lyell was his steady-state geology, we argue in this book that he took much from his steady-state biology as well and that it formed a starting point for his thought in many respects.

Darwin's close connections with ideas emerging from natural theology and uniformitarianism—and presented in our chapter under the rubric of a steady-state world—have been finely analyzed in Levine, G. (1988), *Darwin and the Novelists*, Chicago: University of Chicago Press, 24–55. In our view, however, the *Origin of Species* is so closely tied to these ideas that Darwin could not significantly extricate himself from them, preventing him from initiating what is often called the "Darwinian Revolution." We hold that Darwin created more maneuvering room for evolutionary motion than either Linnaeus and Charles Lyell, but not to the point of moving out of an essentially closed system. The following quote is thus meaningful to us, but not in the way originally intended by its author:

> I have already in the discussion of Whewell tried to suggest how the shadows of the natural-theological argument—particularly its language and its forms—stretch across the pages of the *Origin*. In one sense the theory of natural selection merely provides a different answer to the natural-theological question of *how* organisms adapt. But in the very process words like 'organism' and 'adaptation' and 'species' get redefined by being plunged into history,

quote from Levine, G. (1988), *Darwin and the Novelists*, Chicago: University of Chicago Press, 84. Whereas history definitely plays a role in allowing for evolutionary motion, Darwin restricts its creative action with a long list of explanatory devices presented in our book.

For similarities of views between Darwin and some of his contemporaries concerning the lack of novelties at the class level in the history of life, consult Huxley, T. H. (1862), "The Anniversary Address," *Quarterly Journal of the Geological Society of London*, 18: xl-liv; Owen, R. (1860), "[Darwin on the Origin of Species]," *Edinburgh Review*, 111: 487–532.

The rhetorical dimension of the *Origin of Species* leaves the reader with the impression that life as a whole is characterized by evolutionary change. To take Darwin at his word on this point, however, one must set aside the many passages implicitly or explicitly evoking evolutionary stagnation and equilibrium. In the first edition of the *Origin of Species* (1859), for instance, see pp. 52, 73, 313, 331-2, 351, 368, 390, 400-2, 421, 429, and 488.

For arguments to the effect that Darwin's speciation theory was not very convincing, consult: Romanes, G. (1886) "Physiological Selection: An Additional Suggestion on the Origin of Species," *Journal of the Linnean Society*, 19: 337-411; Vorzimmer, P. (1970), *Charles Darwin: The Years of Controversy*, Philadelphia: Temple University Press; Sulloway, F. (1979), "Geographic Isolation in Darwin's Thinking: The Vicissitudes of a Crucial Idea," *Studies in History of Biology*, 3: 23-65; Mayr, E. (1982), *The Growth of Biological Thought*, Cambridge: Belknap Press, 411-17; and Bulmer, M. (2004), "Did Jenkin's Swamping Argument Invalidate Darwin's Theory of Natural Selection?," *British Journal of the History of Science*, 37: 281-97.

A number of Darwin scholars have argued that the idea of biological progress (or biological complexity) was a central notion for Charles Darwin in the *Origin of Species*. While we claim that Darwin did recognize the necessity of the rise of complexity, we also claim that his theory was entirely powerless to make sense of it. Darwin's support for progress constitutes a merely superficial rhetorical feature of his writing. If one shoehorns all the past under the present, one can only end up with the same variation throughout all historical periods. As seen in previous chapters of this book, Darwin left no explanatory room for the rise of genuine evolutionary novelties. This being said, many Darwin scholars have puzzled over the significance of the idea of biological progress in Darwin's work. Compare Ospovat, D. (1981), *The Development of Darwin's Theory: Natural History, Natural Theology, and Natural Selection, 1838-1859*, Cambridge: Cambridge University Press, 210-11, 225; Mayr, E. (1982), *The Growth of Biological Thought*, Cambridge: Belknap Press, 531-2; Bowler, P. J. (1986), *Theories of Human Evolution: A Century of Debate, 1844-1944*, Baltimore: Johns Hopkins University Press, 41, 150; Ruse, M. (1988), "Molecules to Men: Evolutionary Biology and Thoughts of Progress," in M. Nitecki (ed), *Evolutionary Progress*, 97, Chicago: University of Chicago Press; Hull, D. L. (1988), "Progress in Ideas of Progress," in M. Nitecki (ed), *Evolutionary Progress*, 30, Chicago: University of Chicago Press; Richardson, R. and T. Kane (1988), "Orthogenesis and Evolution in the 19th Century: The Idea of Progress in American Neo-Lamarckism," in M. Nitecki (ed), *Evolutionary Progress*, 151, Chicago: University of Chicago Press; Richards, R. J. (1988), "The Moral Foundation of the Idea of Evolutionary Progress: Darwin, Spencer, and the Neo-Darwinians," in M. Nitecki (ed), *Evolutionary Progress*, 131, Chicago: University of Chicago Press; Gould, S. J. (1996), *Full House: The Spread of Excellence from Plato to Darwin*, Cambridge: Belknap Press, 136–7, 141–2, 144; Ruse, M. (1996), *Monad to Man: The Concept of Progress in Evolutionary Biology*, Cambridge: Harvard University Press, 136–77; Shanahan, T. (2004), *The Evolution of Darwinism: Selection, Adaptation, and Progress in Evolutionary Biology*, Cambridge: Cambridge University Press, 173–95, 285–94; Delisle, R. G. (2019), *Charles Darwin's Incomplete Revolution: The Origin of Species and the Static Worldview*, Switzerland: Springer Nature, 197–228.

For the ongoing debate concerning the "species question" and how Darwin defined species exactly (nominalism versus realism), consult Ghiselin, M. (1969), *The Triumph of the Darwinian Method*, Berkeley: University of California Press, 78–102; Mayr, E. (1982), *The Growth of Biological Thought*, Cambridge: Belknap Press, 265–9; Beatty, J. (1985), "Speaking of Species: Darwin's Strategy," in D. Kohn (ed), *The Darwinian Heritage*, 265–81, New Jersey: Princeton University Press; Stamos, D. (2007), *Darwin and the Nature of Species*, Albany: State University of New York; Sloan, P. (2009), "Originating Species: Darwin on the Species Problem," in M. Ruse and R. J. Richards (eds), *The Cambridge Companion to the Origin of Species*, 67–86, Cambridge: Cambridge University Press; Richards, R. A. (2010), *The Species Problem: A Philosophical Analysis*, Cambridge: Cambridge University Press, 78–96.

Notes

1 Greene, J. C. (1981), *Science, Ideology, and World View: Essays in the History of Evolutionary Ideas*, Berkeley: University of California Press, 12.
2 Young, R. M. (1985), *Darwin's Metaphor: Nature's Place in Victorian Culture*, Cambridge: Cambridge University Press, 144.
3 Hutton, J. (1788), "Theory of the Earth," *Transactions of the Royal Society of Edinburgh*, 1: 304.
4 Young, R. M. (1985), *Darwin's Metaphor: Nature's Place in Victorian Culture*, Cambridge: Cambridge University Press, 136.
5 Huxley, T. H. (1862), "The Anniversary Address," *Quarterly Journal of the Geological Society of London*, 18: xlviii.
6 Owen, R. (1860), "[Darwin on the Origin of Species]," *Edinburgh Review*, 111: 515.

6

Natural Selection: The Core of Darwin's Theory?

Given the panoply of explanatory devices mobilized by Charles Darwin in the *Origin of Species*, it is a bit of a puzzle why they are routinely overlooked by many modern thinkers. It would seem the reasons are multiple—some of which will be discussed in this book's Conclusion—but we will begin with one very modern predilection: our nearly exclusive focus on "natural selection." For many, the name of Darwin is nearly synonymous with the expression of "survival of the fittest": natural selection is imagined as the force primarily responsible for selecting and eliminating variations, which is in turn thought of as a process by which evolutionary change is instituted in biological populations pressed to respond to the adaptive demand of a continually changing environment.

Is this really what the *Origin of Species* is about? Here, the answer must be carefully qualified. It is true that Darwin saw natural selection as the main agent of evolutionary change, and so the inclusion of this expression in the title is to that extent merited: *On the Origin of Species by Means of Natural Selection* (1859). It is at the same time highly misleading, suggesting as it does an account of evolution with natural selection at its center, as if what was on offer was an explanation of evolution *in terms of* natural selection. This is not the case. Natural selection is indeed part of Darwin's theoretical apparatus, but this component of his theory is secondary to other components which take precedence. This may sound like nitpicking, but it is not, for it cuts to the core of common misreadings of the *Origin* and of the misrepresentation of Darwin's place in the history of ideas. Looking at the overall architectonic of the *Origin*, and reading its pages closely, one begins to see that natural selection lies at the periphery of Darwin's theory, not at all at the heart. Darwin's overall view of the history of life does not depend on his understanding of natural selection. Rather, it is erected upon many of the features which are summarized by Darwin himself in his Diagram (see Figure 9) and unpacked in our previous chapters.

This peripheral role of natural selection for Darwin can be seen in the fact that he exploited it only after having put many pieces of his overall picture of evolution in place as a way of accounting for phenomena already observed or recorded. For instance, when Darwin claims that a group of localized islands had been populated by closely related species that originated from a nearby mainland, he postulates that their differences can be explained by each form having encountered different adaptive contexts on separate islands. In this example, which will be further expanded upon below, the notion of natural selection comes during a second phase of the explanatory procedure, merely serving the purpose of *justifying* what had already been established. This illustrates the classic mistake in reading Darwin, which is to invert the logical order of his reasoning and priorities, reinterpreting him according to what we know he *should have thought* rather than what he *did in fact say*. It is precisely because natural selection was not at the theoretical core of his work that Darwin is at liberty to exploit it as freely as he does, claiming it to be at work in some cases, de-emphasizing it or entirely dismissing it in others. When necessary, Darwin even resorts to rather bold speculation in order to make sure that natural selection does not get in his way, as we will demonstrate in the following. This particular feature of Darwin's argumentative strategy will allow us to revisit the rhetorical dimension found in the *Origin of Species*.

Natural Selection Only Comes in after the Fact

As already alluded to, many evolutionists today see themselves as walking in Darwin's footsteps by focusing heavily on the notion of natural selection. For them, "Darwinism" is largely synonymous with natural selection, an evolutionary *process* whose outcome is sometimes reflected in an evolutionary *pattern* known as evolutionary divergence. Under this understanding, it is the evolutionary process that generates the evolutionary pattern. This take on Darwin is summarized by Peter J. Bowler:

> The most radical aspect of Darwin's approach was his reliance on adaptation as the sole driving agent of evolution. Species change because they must adapt to new environments or because they become more specialized for their existing life styles ... Once adaptation is accepted as the sole directing agent, evolution has to be seen as an irregularly branching tree ... If samples from a single original species migrate to several new locations, each will adapt to its new environment in its own way, and eventually a group of distinct but related species will be formed.

> Each of the descendant species will then undergo its own evolutionary process, depending on the opportunities for further migration and adaptation that open up to it. Evolution must thus be regarded as a pattern of haphazard branching, with the branches constantly diverging further apart and redividing where possible ... Darwin devised the theory of natural selection as an explanation of how populations adapt to changes in their environment ... The variations from which the environment selects are essentially random and undirected ... The fact that variations are not directed along fixed lines emphasizes the haphazard or open-ended nature of a process governed only by adaptation.[1]

The reader will easily recognize in this passage the outlines of the *standard view* and the appeal to an evolution of a truly open-ended character—under the influence of natural selection—with *evolutionary contingency* being reflected in haphazard manifestations of local populations adapting to unpredictable conditions. Such was not Darwin's view of evolution, however. This view was in fact developed much later in the twentieth century and retrospectively applied to Darwin by the proponents of the Modern Synthesis, a theory that fuses the causes or processes of evolution with the pattern of evolution. In a recent trenchant critical analysis, Mary P. Winsor rightly stresses:

> The leaders of the Modern Synthesis and their epigones have convinced us that this means that belief in evolution could not be scientific unless it included a cause, which they have achieved by overlooking Darwin's claim that a reasonable person ought to accept branching evolution "even if it were unsupported by other facts or arguments" (Darwin 1859: 458) ... Yet after the Modern Synthesis, many writers treat natural selection and evolution as interdependent and inextricable parts of one unified theory. This flies in the face of the historical facts.[2]

Promoters of the Modern Synthesis and related evolutionary theories take for granted that evolutionary process and pattern are completely unified under a selective theory (adaptation) for which natural selection constitutes the explanatory core. For his part, the actual "unsynthesized" Darwin of the *Origin* treated *process* and *pattern* as two independent explanatory variables that are somehow related but not fused together. The fact that the principle of divergence came first chronologically in Darwin's intellectual development in the 1830s and 1840s may, in part, explain how he came to see pattern and process as separate. Be that as it may, "natural selection" (process) and "divergence" (pattern) are nowhere necessarily linked in the *Origin of Species*. One finds at the heart of Darwin's magnum opus not a pattern explained by the free action of natural selection, but instead a preestablished overall divergent pattern narrowly circumscribing the action of natural selection.

There is more to our story, however, and what follows is perhaps the most interesting part. In order to fully understand why *process* and *pattern* could not have been fused in Darwin's mind, one needs to move beyond a simple opposition between "natural selection" and "divergence." For Darwin, the "principle of divergence" is merely the tip of the iceberg: it is one element in a robust, delicately fleshed out, and consistent view of the history of life—a view organized around a cluster of mutually reinforcing ideas and which owes nothing to natural selection. The notion of divergence does not exist independently of this cluster of ideas; on the contrary, it is only within this cluster that divergence has any meaning or explanatory power. When the notion of natural selection is set alongside this cluster of ideas, it becomes clear why evolutionary *pattern* predominates in Darwin's theory, relegating evolutionary *process* (natural selection) to a secondary role. Pulling together the numerous components of this cluster of ideas already laid out in the previous chapters of this book, this cluster can be summarized in the following way.

The principle of divergence constitutes, for Darwin, a direct application of what we have called his "epistemology of assimilation": as we have seen, past and extinct life forms are subsumed under, or incorporated within, the biological variation of extant or currently living forms. Past forms, therefore, fall entirely in-between the boundaries set by existing ones. This is expressed in Darwin's "cone of increasing diversity" which portrays divergence at a maximum today relative to the entirety of evolutionary history (see Figure 12). There exists no possibility of the past overflowing into the present, of being overtaken by novelties of the past, since the latter are bracketed by the variations of the present. Darwin's biology is, as we have already noted, a neontological science projected backward in geological time. No major evolutionary events are lost to the taxonomic investigation of forms, since the main extant groups are believed to have retained their position relative to one another throughout the entirety of evolutionary history. This explains why Darwin's world has been modern-looking for a very long time, with the Cambrian temporal barrier representing the geological moment when the main extant taxonomic groups arose, nearly fully fledged. When speculating on the other side of that barrier, Darwin resorts to hypothesizing the existence of common ancestors that look like downsized versions of extant forms, pale echoes of the present.

The cluster of ideas just enumerated is supplemented with the "principle of gradation": *Natura non facit saltum* (nature makes no jumps). This principle takes the world to be unified, continuous, and homogeneous, manifesting itself as a network of life forms knit tightly together offering no significant room for evolutionary disruptions. Complementing the "principle of divergence," the principle of gradation ensures that the past stays firmly riveted to the present. Leaving no room for real evolutionary novelties, the homogeneity of the network is characterized by the same rigid and predetermined overall divergent pattern throughout: in the present as in the past, at low taxonomic levels and at high ones.

Finally, and completing our summary, with a long-ago-completed modern world leaving little room for the rise of genuine evolutionary novelties—or not requiring it—it is no surprise that Darwin had little use for a truly *creative* evolutionary dynamics embodied by the process of natural selection. That process is straightjacketed because of the theoretical commitments highlighted earlier: (a) the competitive exclusion principle, involving closely related life forms, tends to exhaust itself during taxonomic ascent as forms grow ever more different from one another; (b) taxonomic groups become ever more stable and entrenched in nature as they climb the taxonomic scale, with little or no change occurring at the upper levels (i.e., classes and phyla); (c) because nearly permanent classes and phyla are rigid and almost eternal, evolutionary motion is suppressed and largely restricted to the cyclical rise and fall of varieties, species, genera, families, and orders; and (d) a significant force in the history of life is that of evolutionary stagnation or stability (equilibrium).

Given Darwin's view of life, evolutionary *pattern* had to take explanatory precedence over evolutionary *process*. Let us now review three scenarios presented by Darwin in which we find him adjusting his strategy from case to case, either channeling, blocking, or entirely dismissing the action of natural selection in order to support his favored evolutionary pattern, one dedicated to placing an overall divergent pattern at the center of the history of life.

Reviewing Three Scenarios

In the first scenario, which concerns the application of the "competitive exclusion principle" previously discussed in Chapter 4, we see Darwin depriving natural selection of its potential creativity by *channeling* its evolutionary dynamics such that it follows a preestablished divergent pattern. Before presenting this example, however, we will first address a possible objection to our example. A cursory reading of the *Origin of Species* might seem to suggest (indeed, has suggested to many) that Darwin sees divergence as a by-product of the action of natural selection, thus fusing both process and pattern into a single explanation, in addition to claiming that the latter is entirely the outcome of the former. Let us revisit our earlier quotation of Darwin discussing this matter:

> But the struggle almost invariably will be most severe between the individuals of the same species, for they frequent the same districts, require the same food, and are exposed to the same dangers. In the case of varieties of the same species, the struggle will generally be almost equally severe … As species of the same genus have usually, though by no means invariably, some similarity in habits and constitution, and always in structure, the struggle will generally be more severe between species of the same genus, when they come into competition with each other … We can dimly see why the competition should be most severe between allied forms, which fill nearly the same place in the economy of nature. (Darwin, 1859: 75–6)

One might want to argue that competition between similar forms creates the conditions for natural selection to favor forms that evolve away from each other, creating a pattern of divergence (1859: 118–19). However, following Darwin one more step in his reasoning reveals a different story. After the passage just cited, Darwin tellingly adds: "[T]he struggle will generally be more severe between species of the same genus … than between species of distinct genera" (p. 76). In other words, when related forms move away from each other—by climbing the taxonomic scale and becoming two distinct varieties, species or genera—they simultaneously become less prone to move away from each other as the conditions prompting the original divergence dissipate (see also Darwin, 1859: 110, 321). This question has been treated more fully in our Chapter 3.

This being the case, there is no reason for Darwin to hold that the overall pattern of evolution is structured around "divergence," as he does in his Diagram (see Figure 9). Here we reach the crux of our argument: Darwin superimposed a pattern of divergence onto the structure of evolution as a whole. He saw this pattern first and then set about explaining it with natural selection. He extrapolated, or took for granted, that such a pattern would persist on the same evolutionary path across all taxonomic levels, repeating itself endlessly. He was already convinced that this was the main feature of evolution, a view of life not based on his understanding of the evolutionary process (natural selection) but rather other presuppositions, some of which we have explained above, and in particular the notion that nature had an in-built pattern or would follow one. Had he begun with natural selection, the resulting overall pattern of evolution— especially at the higher taxonomic levels of families, orders, and classes—would not necessarily be "divergence" but rather, one would expect, a sort of random evolutionary pattern: some lineages would diverge away from each other to various extents; others would draw closer, converging toward each other; others would still keep their relative distance by evolving in parallel or simply remain stable. These various patterns, discussed in our Chapter 1, are in a modern theory of evolution all accounted for in terms of evolutionary contingency. Under this latter understanding, only randomness "governs" the development of living creatures over time. It was Darwin, more than anyone before him, who helped us to arrive at this understanding. Having set the revolutionary wheels in motion, however, he left it to others to trace out all the implications of this view, including just how deeply evolution is characterized by randomness. A shift in understanding of such great magnitude needed time, and many other thinkers, to complete the job.

Let us pursue our investigation a step further. Another scenario sees Darwin *blocking* the forward motion of natural selection. If evolutionary contingency had been a central commitment, Darwin would have treated evolution as an open-ended evolutionary process. Instead, Darwin closed evolution in on itself with extant classes and phyla being envisioned as stable and unassailable entities profoundly entrenched in nature, encountering no serious competition. Within that enclosed world, the role of natural selection is confined to the circular rise and fall of taxonomic entities at lower levels—orders, families, genera, species, and varieties—all channeled into a divergent pattern when evolutionary change is postulated. As we were at pains to show in the last chapter, Charles Darwin's world is both in equilibrium and cyclical. The Darwinian world was compartmentalized in two fundamental ways—vertically and horizontally—creating barriers that effectively disrupt the action of the evolutionary drive (natural selection):

Vertical compartmentalization. Darwin's view of life is not a universal one appealing to a single and unified tree of life. Rather, his evolutionary biology is actively constructed around distinct and separate classes, segregated or segmented entities (1859: 128–30, 207, 450, 483–4). This aspect of his thinking was much in line with his contemporaries and countrymen Thomas Henry Huxley and Richard Owen (as seen in our previous chapter). Darwin's overall view of life, therefore, should be envisioned as a series of independent classes running separately, though side by side, throughout the entire history of life (see Figure 13).

In principle, room for evolutionary change is created when declining forms open up opportunities for ascending ones. Darwin does not contemplate the possibility of classes and phyla going extinct, however. His position arises out of two of his other commitments: (1) That the biological diversity of all known fossils falls in the variational range of extant forms, as seen in the overall divergent pattern of each class since the beginning of life (1859: 329); (2) That evolutionary groups belonging to separate classes have been in existence since the most remote geological periods (1859: 126). For Darwin, then, ever more dominant groups have driven less-dominant ones to extinction, favoring the entrenchment of a limited number of classes and phyla: "[W]e can understand how it is that there exist but very few classes in each main division of the animal and vegetable kingdoms" (1859: 126), or as he states elsewhere, "Thus we can account for the fact that all organisms, recent and extinct, are included under a few great orders, under still fewer classes" (1859: 428–9).

Horizontal compartmentalization. Darwin's world is also seriously lacking in evolutionary opportunities, considering his construal of the Silurian or Cambrian period as a temporal barrier, which divides the world into a "before" and "after" (1859: 302–10, 338; 1872: 285–6): after that barrier, life is modern-looking, organized around the well-defined groups we know today; at the barrier itself, and directly from it, life seems to spring fully fledged; before the barrier one is limited to speculation, hypothesis, and conjecture. In Darwin's view of life, as alluded to previously, the world has long been inhabited by as many different taxonomic groups as encountered today: "[A]t the most remote geological period, the earth may have been as well-populated with many species of many genera, families, orders, and classes, as at the present day" (1859: 126; see also 1872: 97).

In a Darwinian world characterized by the entrenchment and modernity of classes and phyla, the world is also closed in on itself through the relatively unchanging number of species inhabiting it, with the expansion of forms in space and time being roughly proportional to the decline of others (1859: 63–4, 109, 172). Let us revisit this significant passage:

> Thus the appearance of new forms and the disappearance of old forms … are bound together. In certain flourishing groups, the number of new specific forms which have been produced within a given time is probably greater than that of the old forms which have been exterminated; but we know that the number of species has not gone on indefinitely increasing, at least during the later geological periods, so that looking to later times we may believe that the production of new forms has caused the extinction of about the same number of old forms. (Darwin, 1859: 320)

Should we be surprised, then, that Darwin's world is also characterized by the prevalence of evolutionary stability and stagnation (1859: 281, 314–15, 351,

402, 429), as seen in no less than nine lineages out of eleven in his Diagram (see Figure 9)?

Completing this part of our investigation, our final scenario will help illustrate how Darwin downplayed the evolutionary process to ensure that his view of the evolutionary pattern would prevail. Here, we see Darwin *dismissing* natural selection altogether when he needs to by turning to bolder (and suspiciously convenient) explanations. The action of natural selection is thus adjusted at will to preserve a preconceived idea (divergent evolution).

In the scenario concerning the inhabitants of the Galapagos Archipelago (1959: 397–403), Darwin exploits natural selection in a certain way in the first phase of his explanation, only to dismiss it entirely in the second. We will see that this alternating approach came with challenges. According to Darwin, the fauna and flora on these islands 500 or 600 miles off the shores of Ecuador (South America) in the Pacific Ocean share many similarities with the forms found on the nearby mainland. He suggested that these island-dwellers had evolved from common ancestors that originally occupied the mainland, natural selection having adapted them differently to the different conditions encountered on each separate island:

> Thus the several islands of the Galapagos Archipelago are tenanted ... in a quite marvelous manner, by very closely related species; so that the inhabitants of each separate island, though mostly distinct, are related in an incomparably closer degree to each other than to the inhabitants of any other part of the world. And this is just what might have been expected on my view, for the islands are situated so near each other that they would almost certainly receive immigrants from the same original source, or from each other. (Darwin 1859: 400)

Having relied on natural selection as a means to explain the differences between related forms, Darwin now had to face the difficulty of explaining why forms living on specific islands did not typically invade neighboring islands within sight of each other. After all, should not the evolutionary conditions that prompted the colonization of the Galapagos Islands from the mainland also be active within the Archipelago itself? Darwin's surprising answer is a clear "no":

> But we often take, I think, an erroneous view of the probability of closely allied species invading each other's territory, when put into free intercommunication. Undoubtedly if one species has any advantage whatever over another, it will in a very brief time wholly or in part supplant it; but if both are equally well fitted for their own places in nature, both probably will hold their own places and keep separate for almost any length of time ... In the Galapagos Archipelago, many

> even the birds, though so well adapted for flying from island to island, are distinct on each; thus there are three closely-allied species of mocking-thrush, each confined to its own island ... I think we need not greatly marvel at the endemic and representative species, which inhabit the several islands of the Galapagos Archipelago, not having universally spread from island to island. In many other instances, as in the several districts of the same continent, *pre-occupation* has probably played an important part in checking the commingling of species under the same conditions of life. (Darwin 1859: 402–3) [our emphasis]

In this scenario, Darwin shifts his explanatory apparatus to protect the core of his theory, which rests on the notions of common ancestry and divergence, the two being fused in his mind. The issue concerning evolutionary process is thus introduced into the explanation, but only *after the fact*, in order to *justify* this assumption either by using natural selection as a force to split up forms adapted to separate islands or by canceling out this adaptive drive so that they are prevented from successfully invading other islands on the presumption that "pre-occupation" offers each of them an edge.

At this point it might be useful to expand a bit on the rhetorical strategies used by Darwin in the *Origin of Species*. In our Chapter 3, we have seen Darwin resorting to hypothetical explanations in order to fill in blanks of knowledge, as when imagined common ancestors are conjured as substitutes for actual ones. In Chapters 4 and 5, we saw Darwin frequently putting forward seemingly well-supported scientific demonstrations, only to backtrack later with a list of reasons for not accepting them: (a) the assumed ongoing process of divergence across all taxonomic levels versus the gradual exhaustion of the evolutionary drive supporting it; (b) the assumption that "divergence" constitutes the dominant evolutionary pattern in evolution versus life forms being tightly connected in "reproductive networks" preventing divergence; (c) the promotion of biological progress versus manifestations thwarting its expression: stagnation, cyclicity, and equilibrium. Finally, in this chapter we find yet another argumentative strategy exploited by Darwin which, this time, sees him relying on natural selection when convenient for his purposes or dismissing it when this is not the case.

Conclusion

For Darwin, natural selection functions as a handy, all-purpose explanatory component that can be adjusted at will to fit any circumstances, however likely or unlikely they might be. He repeatedly employs this strategy in the service of preserving his theoretical core, which includes the principle of divergence and other components. The action of natural selection is thus subordinated to them, since it can on occasion be channeled, blocked, or entirely dismissed. The standard view that puts natural selection at the center of the *Origin of Species* finds little support on our analysis. Similarly, the now-orthodox process-centered view which envisions natural selection as a central dominant force generating an immense range of possible evolutionary patterns in its wake cannot be found in Darwin: this view is a modern invention, a twentieth-century thought that has been projected backward onto the *Origin*. For Darwin, the notion of evolutionary *pattern* is not linked to the notion of evolutionary *process* as it is for us today. Admittedly, coupling them now seems intuitive, perhaps natural, to us, but this ultimately says more about us than Darwin. It is worth remembering that, forced to choose between natural selection and his overall view of life, Darwin always chooses the latter. Modern evolutionists may perhaps be forgiven for overlooking this, since Darwin was anything but straightforward in his argumentation.

Annotated Bibliography

As throughout this whole book, the two editions of the *Origin of Species* referred to in this chapter are Darwin, C. (1859), *On the Origin of Species*, London: John Murray; Darwin, C. (1872), *The Origin of Species*, 6th edn, London: John Murray.

Among a long list of publications holding the view that natural selection constitutes a core component of Charles Darwin's theory of evolution and Darwinism in general, see Wilson, E. O. (1975), *Sociobiology: The New Synthesis*, Cambridge MA: Harvard University Press; Ruse, M. (1975), "Darwin's Debt to Philosophy: An Examination of the Influence of the Philosophical Ideas of John FW Herschel and William Whewell on the Development of Charles Darwin's Theory of Evolution," *Studies in History and Philosophy of Science*, 6: 159–81; Dawkins, R. (1976), *The Selfish Gene*, Oxford: Oxford University Press; Hodge, M. J. S. (1977), "The Structure and Strategy of Darwin's Long Argument," *British Journal for the History of Science*, 10: 237–46; Mayr, E. (1983), "How to Carry Out the Adaptationist Program?," *American Naturalist*, 121: 324–34; Depew, D. and B. Weber (1995), *Darwinism Evolving*, Cambridge MA: MIT Press; Dennett, D. (1995), *Darwin's Dangerous Idea*, New York: Simon & Schuster; Wray, G., H. Hoekstra, D. Futuyma, R. Lenski, T. Mackay, D. Schluter, and J. Strassmann (2014), "Does Evolutionary Theory Need a Rethink? No, All Is Well," *Nature*, 514: 161–4; Alcock, J. (2017), "The Behavioral Sciences and Sociobiology: A

Darwinian Approach," in R. G. Delisle (ed), *The Darwinian Tradition in Context: Research Programs in Evolutionary Biology*, 37–59, Switzerland: Springer Nature.

Under the standard view, Darwin's theory mainly concerns an evolutionary process (natural selection) whose evolutionary outcome is reflected in a specific pattern (divergence). For different expositions of this view, the reader may consult the following: Browne, J. (1983), *The Secular Ark: Studies in the History of Biogeography*, New Haven: Yale University Press, 215; Bowler, P. J. (1988), *The Non-Darwinian Revolution*, Baltimore: Johns Hopkins University Press, 7–9; Ruse, M. (1999), *The Darwinian Revolution*, 2nd edn, Chicago: University of Chicago Press, Chicago, 191; Kohn, D. (2009), "Darwin's Keystone: The Principle of Divergence," in M. Ruse and R. J. Richards (eds), *The Cambridge Companion to the Origin of Species*, 87–108, Cambridge: Cambridge University Press.

Mary P. Winsor questions the standard view which assumes that evolutionary process and evolutionary pattern are fused in Charles Darwin. See Winsor, M. P. (2009), "Taxonomy Was the Foundation of Darwin's Evolution," *Taxon*, 58: 43–9. We agree with Winsor and go a step further by arguing that the two are so separated in Darwin's mind that evolutionary process (natural selection) only enters the explanation as a mere justification for a preestablished evolutionary pattern (divergence).

During his intellectual development in the 1830s, 1840s, and 1850s, Charles Darwin arrived at the ideas of "divergence" and "natural selection" separately, with the principle of divergence coming first, followed later by the idea of natural selection as a justification for it. For more on the separateness of evolutionary process and evolutionary pattern in Darwin's mind, consult Kohn, D. (1985), "Darwin's Principle of Divergence as Internal Dialogue," in D. Kohn (ed), *The Darwinian Heritage*, 245–57, Princeton: Princeton University Press; Mayr, E. (1985), "Darwin's Five Theories of Evolution," in D. Kohn (ed), *The Darwinian Heritage*, 755–72, Princeton: Princeton University Press; Winsor, M. P. (2009), "Taxonomy Was the Foundation of Darwin's Evolution," *Taxon*, 58: 43–9; Sober, E. (2011), *Did Darwin Write the Origin Backwards? Philosophical Essays on Darwin's Theory*, Amherst: Prometheus Books.

Charles Darwin's view of the history of life has been discussed at some length in Chapters 3, 4, and 5 of this book. The reader is redirected to them for precise pages in the *Origin of Species* wherein Darwin presents the cluster of ideas discussed above.

Robert J. Richards rightly stresses an important fact neglected by most Darwin scholars: Darwin's biology was actively constructed around distinct biological types such as classes. Consult Richards, R. J. (1999), "Darwin's Romantic

Biology: The Foundation of His Evolutionary Ethics," in J. Maienschein and M. Ruse (eds), *Biology and the Foundation of Ethics*, 113–53, Cambridge: Cambridge University Press, see especially p. 129.

Notes

1 Bowler, P. J. (1988), *The Non-Darwinian Revolution*, Baltimore: Johns Hopkins University Press, 7–9.
2 Winsor, M. P. (2009), "Taxonomy Was the Foundation of Darwin's Evolution," *Taxon*, 58: 47.

Part Three

Charles Darwin Viewed in Piecemeal Fashion

7

When So-Called New Ideas Hide Old Ones

We have been asking much of a mid-nineteenth-century scholar such as Charles Darwin. Proponents of the standard view believe he managed to put in place a theoretical framework that, while incomplete in certain respects, provides the foundation upon which later evolutionary theories issued directly. In other words, they argue that Darwin offered the theoretical core of an ongoing research program now known as "Darwinism," which is, in turn, synonymous with modern evolutionary theory. Our analysis does not support such a reading. A brief yet systematic overview comparing the ideas often associated with "Darwinism" will illustrate the significance of the conceptual gap—and incommensurability—between Darwin and evolutionists today.

Evolutionary Contingency

This key notion encountered many times in our journey holds that evolution is an open-ended process. It must be, since the overall picture consists in a score of independent lineages finding as many different adaptive solutions as they can, on the basis of the set of variations each carries at a particular time and place in the history of life. The meeting of unpredictable biotic and abiotic changes creates the awesome spectacle we see everywhere: life forms evolving in all sorts of evolutionary directions, just as we would expect given the opportunistic nature of life (as we now understand it). However, we have seen that by emphasizing divergence in the way that he did, Darwin put evolution into a straightjacket, effectively closing off a number of significant evolutionary options to lineages, such as evolutionary convergence and reticulate evolution. Evolutionary contingency is a major conceptual innovation, which, after much time and thought, ultimately superseded *predeterminism*, a major leap of thought that could not be contemplated by scholars like Darwin who, in the nineteenth century, were still struggling to draw out the full range of implications accompanying the intellectual transition between static and evolutionary worldviews. Evolutionary contingency began to be more fully grasped only during the twentieth century, when evolutionary biology and our understanding of our place in nature had reached a greater level of maturity. A significant degree of determinism and predictability persists in Darwin's view of life, in contradistinction to our modern view.

Natural Selection

In a truly open-ended world, natural selection makes for the central explanatory component for organizing evolution. Indeed, natural selection becomes the "decider" of the encounter between the variations carried by life forms and the environmental contexts they face—the outcomes being either positive selection (and possible evolutionary change) or negative selection (and eventual extinction). A theory devoted to explaining such a world has *adaptation* at its main focus. We have seen that Charles Darwin did not put natural selection at the explanatory core of his theory. For him, evolutionary pattern has explanatory precedence over evolutionary process. That is why natural selection merely plays the role of justifying Darwin's view of life, needing to be either channeled, blocked, or dismissed, as circumstances require. The theory promoted by a number of modern evolutionists is not Darwin's theory: the former should not be called "Darwinian" but more appropriately, perhaps, "selective theory."

Adaptation

We would expect a selective evolutionary theory founded on a truly open-ended set of processes to generate all kinds of adaptations, including the appearance of similar traits in forms not particularly closely related (owing to evolutionary parallelism or convergence) and the deletion of traits in related forms (following adaptive change). In both cases, the process of adaptation blurs to a greater or lesser extent the exact nature of the relationship between life forms. Charles Darwin refused to commit himself to this kind of *adaptationist approach*. Instead, he thought he had a fairly clear view of the evolutionary past of the main extant groups under the assumption that they kept their position relative to one another throughout the entirety of evolutionary history. As already seen, the phylogenetic positioning of all main taxonomic groups is now known to us, Darwin (1859: 420–2; 1872: 369–70) assumes. This would imply that natural selection has not succeeded in erasing traces left by genealogical connections, nor has it given rise to similar yet unrelated traits, to any significant extent. Darwin relied on his view of life—which emphasized the principles of divergence and gradation, among other ideas—to support this commitment.

The Principle of Divergence and Common Ancestry

In the context of a truly open evolutionary process, the notions of "divergence" and "common ancestry" ought to be largely independent of one another. Having emerged out of a common ancestor, any two lineages are free to evolve in all sorts of directions, subject to any of the patterns laid out in our earlier chapters—divergence, convergence, reticulate evolution, stagnation, and extinction—depending of course on the specifics of the environmental contexts they each face (biotic and abiotic) as well as on the biological variations they separately carry (see Figure 7). For his part, Darwin treated "common ancestry" and "divergence" as two expressions of the same general idea: "the cone of increasing diversity" (see Figure 12), the idea that all past and extinct forms fall in-between extant ones. When evolution occurs, Darwin holds that change is channeled into a single pattern, that of divergence. Two related forms stemming from the same ancestor have no other option, insists Darwin, but to diverge away from each other (unless one or both remain stable or go extinct). The fusion of the notions of divergence and common ancestry at the core of Darwin's theory deprive the evolutionary process of the adaptive flexibility our modern acceptance of evolutionary contingency requires.

A Self-regulated Economy of Nature

When nature is envisioned through the lens of evolutionary contingency, the proportions between its constituent parts—the numerous life forms—are in no way predetermined. Rather, these proportions are freely established or *self-regulated* by the confrontation of the various actors with each other and their surroundings at each moment in the history of life. Life forms having the right variations in the proper context will thrive; all others will struggle and ultimately fail. Given his view as whole, Darwin could not have embraced this aspect of the modern view. His overall account of the rolling out of life over time imposes constraints on the free play of the wild, untrammeled forces central to our modern theory. He explains evolution using several devices that of their nature serve to reign in this free play: (1) channeling evolution under an overall divergent outlook; (2) favoring evolutionary equilibrium or stagnation; (3) entrenching taxonomic entities at higher taxonomic levels (i.e., classes and phyla); (4) allowing for premature exhaustion of the evolutionary drive under the competitive exclusion principle as species ascend the taxonomic scale; (5) taking for granted that the overall number of species has remained constant throughout the history of life; (6) promoting the cyclicity of rising and declining entities at lower taxonomic levels: varieties, species, genera, families, and orders, and the narrowing of evolutionary motion within these confines.

Biological Variation

Finally, a truly open evolutionary process implies that biological variation generated during the course of evolution is, in principle, open to an indefinite number of manifestations. Given enough geological time, and a range of species, along with fluctuating biotic and abiotic contexts, conditions will allow for the rise of a massive amount of biological diversity. This, again, is not the approach taken by Darwin. Projecting the variability of extant forms backward into geological times, he closed evolution in on itself by merely recycling the same variations through time. It is no surprise that his so-called common ancestors or progenitors look suspiciously like downsized forms awaiting their ultimate and full expression only later in their modern representatives. We have seen that Darwin's way of addressing this issue is akin to the doctrine of preformationism.

Conclusion

Charles Darwin was no modern evolutionist. The list of intellectual commitments he inherited from the static worldview is long and reveals the many respects in which he belongs to an earlier time. Darwin is more appropriately seen as attempting to establish a theory of evolution by stretching the conceptual envelope of the prevailing views of earlier centuries, in particular the static worldview, rather than founding a wholly new approach closely resembling our current conceptions in its essentials. In short, he should be seen as traveling the end of an old road than the beginning of a new one. The establishment of a genuinely selective theory based on evolutionary contingency required much more than just tweaking Darwin's original proposals: it required a conceptual breakthrough from the previous ages.

Of course, we are not denying the obvious contribution Charles Darwin made to our modern conception of evolution, through his words and his example. However great, it was still *a single* contribution, and on that we must insist, for several reasons. First, there is the matter of historical accuracy: as we have been at pains to show in the preceding pages, Darwin simply did not hold, and *could not have* held, the beliefs ascribed to him by modern scholars. We moderns have been all too eager to find ourselves in Darwin and have "succeeded" by force of will, at the price of overlooking our profound originalities. In addition to getting Darwin and his contributions wrong, collapsing Darwin's thought into our own further muddies the waters in an already murky area of human thought. To truly understand evolution, one must see the workings of each of its components clearly, as well as their interrelationships. This is something we have become much better at than Darwin ever could have been, given his starting point and available evidence. Only confusion can spring from claiming that he had things largely figured out from the start (and such claims are far from rare, as some of the quotations in our "Introduction" attest). Moreover, in the area of science, mistakes often count as important results. To deny Darwin his mistakes is to deny ourselves knowledge of a kind.

Second, for many, Darwin's thought and its relation to his own times are interesting in their own right, not merely as steps along the way reaching a pinnacle today, a vehicle handily transporting us to our current views. Here again, there is great value in getting Darwin right on his own terms, not merely in terms of his stretch of a trajectory now seen as complete. There is an inherent danger of imposing the equivalent of a "Whiggish" notion of history here—one in which all past events are seen as having happened in direct service of the present.

Finally, no one benefits from mythmaking or oversimplification when the matter under consideration is science or its history, nor do they profit from the lionizing of individual isolated figures. This is not true to the facts, either with respect to the hard and slow (and nearly always collaborative) work of science, nor to the modification of thought more generally. Darwin's influence was diffuse, and his thought was absorbed by a community of thinkers in a *piecemeal fashion*. It took time to digest, to gestate, and to germinate new thoughts, by many men and women working over many years. This is evidenced in the many research programs in evolutionary biology that have appeared since the twentieth century, which have differentially and profoundly *reconceptualized* many of Darwin's notions, employing them at times under competing, even incompatible, perspectives. It is a testament to Darwin's influence and the fruitfulness of his ideas that a bit of him can be found in a wide range of research programs, but this also means none of them can justifiably claim Darwin as uniquely their own. And this is exactly as we should expect given the approach to Darwin we are commending: this way of seeing the development of ideas—diffuse and growing slowly out of what has come before—stands in sharp contrast to the "founding figure," bolt-of-inspiration accounts many of us were offered in school, and which still prevail among scholars and lay people alike. It also resonates with Darwin's oft-cited motto: *natura non facit saltum*. While this principle has proven to be less than universally valid with respect to evolution, it is still not a bad place to start when considering the development of ideas.

Annotated Bibliography

Throughout the *Origin of Species*, Charles Darwin struggled with the possibility that his neat divergent outlook would be blurred or confused by alternatives like evolutionary convergence and reticulate evolution. Because he wanted to keep the past under the control of the present—keeping in mind what we have here referred to as his *epistemology of assimilation* in which extinct forms fall-in between extant ones—such alternatives presented themselves as genuine threats to his theoretical core, which included the principle of divergence. When confronted with evolution's openness and complexities, Darwin largely obscured that reality by imposing his own agenda upon it. Consult Delisle, R. G. (2019), *Charles Darwin's Incomplete Revolution: The Origin of Species and the Static Worldview*, Switzerland: Springer Nature, 85–105, 161–76.

Ernst Mayr has argued that in Darwin the notions of "divergence" and "common ancestry" are two separate theories. Consult Mayr, E. (1985), "Darwin's Five Theories of Evolution," in D. Kohn (ed), *The Darwinian Heritage*, 755–72, Princeton: Princeton University Press. Mayr's thesis is not supported by our analysis, however. In the *Origin of Species*, Darwin entirely blends the two notions on the understanding that evolution is a preestablished divergent pattern. It cannot be otherwise if one claims that all extinct forms fall in-between extant forms.

Ernst Mayr is also responsible for popularizing the thesis that Charles Darwin was the founder of a revolutionary shift in thinking by initiating what Mayr calls "population thinking." This way of seeing things takes differences or biological diversity between individuals within a population to be a keystone of evolution, as natural selection seizes upon such differences to institute evolutionary change or adaptation. Mayr suggests that Darwin's populational approach contrasts with "typological thinking," the notion that differences between organisms should be dismissed in order to favor more inclusive biological types or essences. Consult Mayr, E. (1963), *Animal Species and Evolution*, Cambridge MA: Belknap Press, 5–6; Mayr, E. (1982), *The Growth of Biological Thought*, Cambridge MA: Belknap Press, 45–7; Mayr, E. (1991), *One Long Argument: Charles Darwin and the Genesis of Modern Evolutionary Thought*, Cambridge MA: Harvard University Press, 40–2. In light of our analysis, Mayr's thesis loses a significant part of his argumentative force. By locking up evolution inside entrenched extant classes and phyla, Darwin embraced a form of typological thinking.

We hold that Charles Darwin cannot be seen as a foundational figure of modern evolutionism. Rather, his influence on so-called neo-Darwinism or the Modern Synthesis is much weaker than is usually believed. Yet, at the same time, his influence was more diffuse and widespread than usually argued, as a number of his proposed notions were differentially reconceptualized and inserted in a number of competing research programs in the twentieth and twenty-first centuries. Consult Levit, G. S. and U. Hossfeld (2011), "Darwin without Borders? Looking at Generalised Darwinism through the Prism of the Hourglass Model," *Theory in Biosciences*, 130: 299–312; Delisle, R. G. (2017), "Introduction: Darwinism or a Kaleidoscope of Research Programs and Ideas?," in R. G. Delisle (ed), *The Darwinian Tradition in Context: Research Programs in Evolutionary Biology*, 1–8, Switzerland: Springer Nature; and Delisle, R. G. (2017), "From Charles Darwin to the Evolutionary Synthesis: Weak and Diffused Connections Only," in R. G. Delisle (ed), *The Darwinian Tradition in Context: Research Programs in Evolutionary Biology*, 133–67, Switzerland: Springer Nature.

Conclusion: Back to the Future

This book dared to ask a surprising (and at first glance, perhaps even preposterous) question: "Was Charles Darwin truly an evolutionist?" We have seen that merely asking the question dislodges many overlooked assumptions about Darwin. In many respects, the work of the *Origin of Species* consisted in pushing the conceptual envelope of the static worldview to its limit in the context of a rising evolutionism. When these more antiquated aspects of Darwin's thought are blended with the more modern of them, a picture begins to emerge of a scholar in transition between two worldviews, the static and the evolutionary. Charles Darwin was an evolutionist, to be sure, but of a very special kind, one far removed from the mythical status awarded him as the prophet and forerunner of this new movement. The rise of evolutionism in the eighteenth and nineteenth centuries involved a long and challenging shift of perspective with respect to how we see our world, one that has been accompanied less by great leaps and more by all sorts of compromises between the old and the new. Darwin was no exception in this respect, and offers us one such fascinating compromise.

If, as we have claimed, Darwin had both feet planted firmly in the past, why have scholars been reluctant to register the extent to which Darwin was committed to the older worldview? An adequate response to this question would require a book-length investigation, but we will hazard a few tentative suggestions, some of which are more controversial than others. First, the explanations provided in the *Origin of Species* are filled with profound contradictions, which are partly concealed by Darwin's rhetorical skill in smoothing them over by means of his unifying discourse. Although evolutionists of the twentieth and twenty-first centuries have largely overlooked the most important of these contradictions, it is interesting to note that a contemporary, as well as a friend and supporter of Darwin, noticed that there might be a problem lurking here. In a book review following the publication of the first edition of the *Origin*, T. H. Huxley wrote in 1860:

But this very superabundance of matter must have been embarrassing to a writer who, for the present, can only put forward an abstract of his views; and thence it arises, perhaps, that notwithstanding the clearness of the style, those who attempt fairly to digest the book find much of it a sort of intellectual pemmican—a mass of facts crushed and pounded into shape, rather than held together by the ordinary medium of an obvious logical bond: *due attention* will, without doubt, discover this bond, but it is often hard to find.[1] [our emphasis]

With friends like this, one might say, one hardly needs any enemies! Unfortunately, "due attention" has only revealed more profound ambiguities still.

Second, one prevailing view of science sees the enterprise as primarily devoted to producing and elaborating "theories." A *theory* is a compact explanatory structure organized around a coherent group of principles offered to explain natural phenomena. For instance, Isaac Newton's theory of universal gravitation explains phenomena like falling objects, cycles of tides, and revolving planets; it is a canonical illustration of what good theories can achieve. It is difficult to resist the appeal of the explanatory power such theories provide. Part of it, too, lies in the aesthetic qualities of the "theory-view." It is precisely this view of science that has been retrospectively applied to the *Origin of Species* over the past decades by scholars of different stripes, overlooking its internal contradictions in favor of the ostensible coherence of a unified "theory" subsuming a wide range of distinct evolutionary phenomena under a single umbrella. They went looking for it and, unsurprisingly, they found it.

It has become increasingly apparent that the coherence of the *Origin* is illusory, largely a by-product of a "theory-view" of science imposed on it from without. The quest for an evolutionary biology structured as a unified theory has largely eluded us to this day, likely owing to the fact that it is an area of the natural world that is just too complex to be reducible to a single theory or even a cluster of tightly connected theories. Indeed, our quest for unified knowledge has thus far been stymied by the wealth of disparate facts about living phenomena to be found in the living world. It is likely that this area will require applying a more complex and flexible view of science based not on theories per se, but rather on "research traditions" or "worldviews." This way of understanding the march of science allows for profound changes in perspective over time and even for the formulation of contradictory approaches within their confines.

Third, a lack of critical analyses by historians and philosophers further contributed to distorting what Charles Darwin had really achieved. Indeed, a limited number of influential biologists have proclaimed that they arrived at a unification of knowledge during the 1930–60 period under the appellation of the "Modern Synthesis." That synthesis, they claimed, built its explanatory edifice upon what they took to be Darwin's own theoretical core (i.e., natural selection) in reaching the conclusion that evolution is a gradual process operating under the action of natural selection playing on small (genetic and organismic) variations. Strangely enough, few historians or philosophers objected to this self-justificatory discourse concerning unification in evolutionary biology. Instead, the former simply became promoters of the latter's discourse. It seems that early on (beginning in the 1960s) historians and philosophers of biology had relinquished the measured skepticism their profession required. As was to be expected, work recently undertaken employing a more standard, critical approach has revealed that the longed-for unification of knowledge in biology can hardly be taken for granted, whether in Darwin's case or in that of the so-called Modern Synthesis, as well as in other attempts at a synthesis made since. History is too important and complex to be left to those who have an interest in simplifying it for self-serving ends.

Fourth, Charles Darwin's work has become caught up in the crossfire of ideological questions. For complex reasons, Darwin became the solitary shining symbol of evolution in the twentieth and twenty-first centuries. This was especially obvious during the 100th anniversary of the *Origin of Species* (1859), which was celebrated with great pomp and fanfare in 1959. As summarized by Betty Smocovitis in 1999:

> Darwin and his life and work held powerful symbolic meaning for postwar evolutionary biologists, who were eager to unify, strengthen, and promote their new-found community ... [The] celebration was not the exclusive domain of scientists and academics ... The rather tepid, nontechnical scientific discussions were intended for mostly general audiences ... [Evolution] despite the continuing controversy over science and religion, was a recognizable fact. With proper knowledge of this evolutionary past, humans would be able to control their evolutionary future. Evolutionary progress and social progress were thus inextricably linked for American popular audiences; their own increasingly technological culture had, after all, progressively evolved ... This celebration ... had little to do with the historical Darwin or the development of his work; instead it revealed much about postwar American culture.[2]

Whether or not one accepts the particulars of Smocovitis's analysis, it reveals how tightly scientific concerns are intertwined with sociological forces. The sesquicentennial celebrations of the *Origin* in 2009 were no less extravagant, with a score of events and publications. In this context, a prime occasion for myth-building, it is no surprise that evolution's enemies targeted Darwin in particular, wrongly assuming that proving him wrong on this or that point would shake the edifice of evolution itself, as if unaware that Darwin was neither the first, last, or only evolutionist. One person's "Great Man," it seems, is all-too-easily transfigured into a "Straw Man." A troubling aspect of this story can be seen in the reactions of some of the evolutionists themselves, however: in order to defend evolutionary science against creationist attacks, they responded by placing Darwin on a pedestal, thus concentrating attention on him as an individual. Darwin was turned into a symbol, a hero, a saint of science, whose reputation would act as a rampart against fundamentalist attacks and preserve the luster of science in the eyes of nonscientists. In ways subtle and unsubtle, this has placed Darwin above criticism, even among scholars. To answer ideological dogmatism with scientistic dogmatism is perhaps the worst strategy that could have been adopted.

What separates science from nonscience is precisely the freedom of scholars and researchers to evaluate all claims critically, past and present. A healthy and progressive science must be a critical science, always reassessing its own views and findings, and this is equally true of history. The image of Charles Darwin cannot be fixed for all eternity in the name of an ideological war, foreclosing legitimate investigations about his work. What's more, the strategy of using Darwin as the vanguard in the fight against evolution's opponents will ultimately be less effective than forcing them to confront the overwhelming array of evolutionary phenomena, ideas, and thinkers who despite substantial differences converge on the *fact* of evolution. Acknowledging Darwin's intellectual indebtedness to the past, in particular to his grounding in the static worldview, may mean Darwin is no longer in a privileged position in the ideological combat against religion, but this is a rather small price to pay: he and his work are of immeasurably greater value as a thinker than as a hero, villain, or stalking horse.

Finally, the long overdue reassessment of our understanding of Charles Darwin's work has probably been delayed by the professional training of evolutionists themselves. The field of evolutionary biology encompasses an array of disciplines—molecular biology, genetics, ethology, morphology, ecology, zoology, paleontology, embryology—which, though strikingly different in many respects, share a focus on studying extant or currently existing forms. Paleontology excepted, their shared focus on these forms means they have a distinctly neontological orientation. Darwin failed to complete his transition to modern evolutionism in large part because he did not take stock of the profound historicity of life. Combating a very specific set of ideas, he thought it would suffice to collect the proof in favor of evolution from within the present time horizon, merely interpolating that knowledge into the past and extrapolating it into the future. As we have seen, however, his attempt at understanding evolutionary change with such an approach left little room for the recognition of the rise of genuine novelties, depriving evolution of creative time.

Neontologists today are similarly unable to recognize Darwin's explanatory weaknesses, we argue, for the simple reason that they still suffer from his same limitations, at least to a degree. For too many of them, it seems, evolutionary biology is centered around evolutionary mechanisms as they appear today and projected in the time dimension, reading the past through the lens of the present. It would thus appear that the field of evolutionary biology has yet to reach its full maturity, insofar as it has not managed to integrate the profound historicity of the world into its investigative approach. The transition between the static and evolutionary worldviews is, indeed, a long and difficult one, still very much in progress.

Things have improved recently with the rise of the field of "paleobiology," which, at last, takes seriously information extracted from the geological past, although many of its ideas, facts, and methods have yet to germinate throughout evolutionary biology more generally. This contextualizes and helps calibrate our remarks throughout this book and should alleviate the potential worry that we have been too harsh in our criticisms: one can hardly blame Darwin for not having successfully accomplished a transition we are still struggling with to this very day. It also shows the value of taking him as thinker rather than a prophet, since pulling him down from his pedestal returns attention to his accomplishments rather than his failings. The investigation of this book has therefore come full circle: what started as an inquiry concerned with Darwin's approach to the past ends by raising the same concern regarding evolutionists today.

Revisiting the *Origin of Species* opens the door to understanding much more than Charles Darwin's actual words. It is also an encounter with critical questions facing civilization: What is the nature of science? How should science be carried out and studied? What is the role played by "mythical" figures in science and history? How should scientists engage with religious dogmas and ideological issues? We will not hazard answers to these questions here; our purpose, as alluded to in our Introduction, is simply to clear out sufficient space for them to be asked and to insist that we keep asking them. Beyond that, this book carries no agenda whatsoever, hidden or otherwise. It is not intended as a diatribe against Charles Darwin but, we freely admit, addresses a certain kind of suffocating Darwin-worship. It is written in the name and spirit of good science, the pursuit of which requires a searching and critical history, whether this is the history of ideas or the history of life itself.

Annotated Bibliography

Again, though Charles Darwin became intrigued by the idea of evolution while still young, he was not an evolutionist at that time. He later began an intellectual process that saw him edging slowly away from the static worldview in its original form. By the time he published the *Origin of Species* in 1859, Darwin had reached the intermediary point described in our book. To get a glimpse of how far along he had traveled on this intellectual path by 1844, consult Partridge, D. (2018), "Darwin's Two Theories, 1844 and 1859," *Journal of the History of Biology*, 51: 563–92. What Partridge says about the *Origin of Species* (1859) in his paper is, however, not shared by us since based on the standard view.

T. H. Huxley's review of the *Origin of Species* is found in Huxley, T. H. (1860), "Darwin on the Origin of Species," *Westminster Review*, 17: 541–70. For an analysis of the reviews of Charles Darwin's work among his contemporaries, see Ellegard, A. ([1958] 1990), *Darwin and the General Reader: The Reception of Darwin's Theory of Evolution in the British Periodical Press, 1859–1872*, Chicago: University of Chicago Press.

For a rapprochement between the Newtonian synthesis and the so-called Modern Synthesis, see Holton, G. (1978), *The Scientific Imagination: Case Studies*, Cambridge: Cambridge University Press, 11–151. For the connection between Newton, Darwin, and the Modern Synthesis, consult Smocovitis, V. B. (1996), *Unifying Biology: The Evolutionary Synthesis and Evolutionary Biology*, Princeton: Princeton University Press, 170–1.

The inquiry into the feasibility of erecting a Newtonian-style theory in evolutionary biology or within Darwinism itself intersects at points with the issue of deductive-like versus inductive-like explanations in science. Consult Woodger, J. (1948), *Biological Principles*, London: Routledge & Kegan Paul; Beckner, M. (1959), *The Biological Way of Thought*, New York: Columbia University Press; Williams, M. (1970), "Deducing the Consequences of Evolution: A Mathematical Model," *Journal of Theoretical Biology*, 29: 343–85; Ruse, M. (1973), *The Philosophy of Biology*, London: Hutchinson University Library; Hull, D. L. (1973), *Darwin and His Critics*, Cambridge: Harvard University Press, 16–36; Hodge, M. J. S. (1977), "The Structure and Strategy of Darwin's Long Argument," *British Journal for the History of Science*, 10: 237–46; Caplan, A. (1978), "Testability, Disreputability, and the Structure of the Modern Synthetic Theory of Evolution," *Erkenntnis*, 13: 261–78; Ruse, M. (1979), *The Darwinian Revolution*, Chicago: University of Chicago Press, 56–63, 174–80, 197–8, 234–9; Tuomi, J. (1981), "Structure and Dynamics of Darwinian Evolutionary Theory," *Systematic Zoology*, 30: 22–31; Williams, M. (1982), "The Importance of Prediction Testing in Evolutionary Biology," *Erkenntnis*, 17: 291–306; Hodge, M. J. S. (1989), "Darwin's Theory and Darwin's Arguments," in M. Ruse (ed), *What the Philosophy of Biology Is*, 163–82, Dordrecht: Kluwer; Ruse, M. (2000), "Darwin and the Philosophers: Epistemological Factors in the Development and Reception of the Theory of the *Origin of Species*," in R. Creath and J. Maienschein (eds), *Biology and Epistemology*, 3–26, Cambridge: Cambridge University Press; Ruse, M. (2009), "The History of Evolutionary Thought," in M. Ruse and J. Travis (eds), *Evolution: The First Four Billion Years*, 1–48, Cambridge: Belknap Press.

For classic expositions of the nature of scientific **"theories"**, see Hempel, C. (1966), *Philosophy of Natural Science*, New Jersey: Prentice Hall; Nagel, E. (1979), *The Structure of Science: Problems in the Logic of Scientific Explanation*, Indianapolis: Hackett. Moving beyond a "theory-view" of science, the nature of science can be viewed through more complex and flexible explanatory structures, often loosely described as "conceptual frameworks": (1) **"Research programmes"** focus on an explanatory core protected by surrounding components (see Lakatos, I. [1970], "Falsification and the Methodology of Scientific Research Programmes," in I. Lakatos and A. Musgrave (eds), *Criticism and the Growth of Knowledge*, 91–196, Cambridge: Cambridge University Press). What was presented in our Chapter 6 can profitably be examined through the lens of Lakatos's "research programmes" (see Delisle, R. G. [2021], "Natural Selection as a Mere Auxiliary Hypothesis (Sensu Stricto I. Lakatos) in Charles Darwin's *Origin of Species*," in R. G. Delisle (ed), *Natural Selection: Revisiting Its Explanatory Role in Evolutionary Biology*, 73–104, Switzerland: Springer Nature); (2) **"Paradigms"** assume that empirical facts, theoretical issues, and methodological choices are all intimately related with one another, even to the exclusion of other competing paradigms (see Kuhn, T. [1970], *The Structure of Scientific Revolutions*, 2nd edn, Chicago: University of Chicago Press); (3) **"Research traditions"** combine what has been said of both research programs and paradigms, although *research traditions* are conceived as entities that can profoundly change over time in addition to being more open structures by allowing for the elaboration of contradictory theories within their confines (see Laudan, L. [1977], *Progress and Its Problems: Towards a Theory of Scientific Growth*, Berkeley: University of California Press); (4) **"Worldviews"** have been illustrated in this book through Charles Darwin's commitment to the static worldview and consist in research entities that can be conceptually stretched to a considerable extent (see Greene, J. C. [1981], *Science, Ideology, and World View: Essays in the History of Evolutionary Ideas*, Berkeley: University of California Press and DeWitt, R. [2010], *Worldviews: An Introduction to the History and Philosophy of Science*, 2nd edn, Oxford: Wiley-Blackwell). It should be noted that it is not always easy to clearly distinguish between "theories," "research programmes," "paradigms," "research traditions," and "worldviews" as these notions sometimes shade into one another and are to some extent a matter of definition. For a very useful and systematic comparison of the views of philosophers of science such as T. S. Kuhn, I. Lakatos, L. Laudan, and P. Feyerabend, consult Laudan, L., A. Donovan, R. Laudan, P. Barker, H. Brown, J. Leplin, P. Thagard, and S. Wykstra (1986), "Scientific Change: Philosophical Models and Historical Research," *Synthese*, 69: 141–223.

For self-assessed declarations by some influential biologists to the effect that they have reached a unification of knowledge in evolutionary biology during the 1930–60 period, see Huxley, J. S. (1942), *Evolution: The Modern Synthesis*, London: Allen and Unwin; see especially the preface; Simpson, G. G. (1949), *The Meaning of Evolution*, New Haven: Yale University, 277–9; Dobzhansky, T. (1949), "Toward a Modern Synthesis," *Evolution*, 3: 376–7; Rensch, B. (1980), "Historical Development of the Present Synthetic Neo-Darwinism in Germany," in E. Mayr and W. Provine (eds), *The Evolutionary Synthesis: Perspectives on the Unification of Biology*, 284–303, Cambridge: Harvard University Press; Mayr, E. (1982), *The Growth of Biological Thought*, Cambridge: Belknap Press, 566–70.

Under the somewhat self-congratulatory rubric of the "Modern Synthesis" that allegedly took place between 1930 and 1960, these same biologists turned to Darwin, declaring their Modern Synthesis to be an extension of Darwin's work. A few statements of this presumed affinity include (1) Theodosius Dobzhansky's book mimicking the title of the *Origin of Species* (Dobzhansky, T. [1937], *Genetics and the Origin of Species*, New York: Columbia University Press); (2) George Gaylord Simpson's essay entitled "One Hundred Years without Darwin Are Enough" (Simpson, G. G. [1964], *This View of Life*, New York: Harcourt, Brace & World, 26–41); and (3) Ernst Mayr's 1991 book entitled *One Long Argument: Charles Darwin and the Genesis of Modern Evolutionary Thought*, Cambridge MA: Harvard University Press.

It is perfectly legitimate and desirable for historians and philosophers to interact with biologists for the benefit of our understanding of evolutionary biology. But a sane and critical distance must be kept by the former at all times to avoid being co-opted by biologists, lest they become the handmaidens of science. A progressive science can only be a critical one. For the work of some historians and philosophers who, in our view, have uncritically embraced the self-justificatory discourse of biologists promoting the Modern Synthesis between 1930 and 1960 (including its Darwinian roots), see Ruse, M. (1973), *The Philosophy of Biology*, London: Hutchinson University Library; Hodge, M. J. S. (1977), "The Structure and Strategy of Darwin's Long Argument," *British Journal for the History of Science*, 10: 237–46; Provine, W. B. (1978), "The Role of Mathematical Population Geneticists in the Evolutionary Synthesis of the 1930s and 1940s," *Studies in History of Biology*, 2: 167–92; Ruse, M. (1979), *The Darwinian Revolution*, Chicago: University of Chicago Press; Provine, W. B. (1985), "Adaptation and Mechanisms of Evolution after Darwin: A Study in Persistent Controversies," in D. Kohn (ed), *The Darwinian Heritage*, 825–66, Princeton: Princeton University Press; Provine, W. B. (1988), "Progress in Evolution and Meaning in Life," in M. N. Nitecki (ed), *Evolutionary Progress*, 49–74, Chicago: University of Chicago Press. This list of scholars could easily be expanded. We are not contesting the quality of their work here. We are, however, pointing out that significant distortions are introduced into our understanding of the development of science when historians and philosophers confound their interests with those of practicing scientists.

In the competitive context of limited institutional resources, it is perfectly understandable that some biologists would wish to present a unified front to the outer world. Unfortunately for those biologists, the sociological needs driving such a unification have not been accompanied by a deeper unification at the conceptual level. This is slowly being revealed in a series of recent analyses. Consult Levit, G. S., M. Simunek, and U. Hossfeld (2008), "Psychoontogeny and Psychophylogeny: Bernhard Rensch's (1900–1990) Selectionist Turn through the Prism of Panpsychistic Identism," *Theory in Biosciences*, 127: 297–322; Delisle, R. G. (2008), "Expanding the Framework of the Holism/Reductionism Debate in Neo-Darwinism: The Case of Theodosius Dobzkansky and Bernhard Rensch," *History and Philosophy of the Life Sciences*, 30: 207–26; Cain, J. (2009), "Rethinking the Synthesis Period in Evolutionary Studies," *Journal of the History of Biology*, 42: 621–48; Van Der Meer, J. M. (2009), "Theodosius Dobzhansky: Nothing in Evolution Makes Sense Except in Light of Religion," in N. Rupke (ed), *Eminent Lives in Twentieth-Century Science and Religion*, 2nd edn, 105–27, Oxford: Peter Lang; Delisle, R. G. (2011), "What Was Really Synthesized during the Evolutionary Synthesis? A Historiographic Proposal," *Studies in History and Philosophy of Biological and Biomedical Sciences*, 42: 50–9; Levit, G. S. and U. Hossfeld (2011), "Darwin without Borders? Looking at Generalised

Darwinism through the Prism of the Hourglass Model" *Theory in Biosciences*, 130: 299–312; Delisle, R. G. (2017), "From Charles Darwin to the Evolutionary Synthesis: Weak and Diffused Connections Only," in R. G. Delisle (ed), *The Darwinian Tradition in Context: Research Programs in Evolutionary Biology*, 133–67, Switzerland: Springer Nature; Adams, M. B. (2021), "Little Evolution, Big Evolution: Rethinking the History of Darwinism, Population Genetics, and the Synthesis," in R. G. Delisle (ed), *Natural Selection: Revisiting Its Explanatory Role in Evolutionary Biology*, 195–229, Switzerland: Springer Nature.

It is interesting to note that Ernst Mayr, a prominent architect of the so-called Modern Synthesis, had this to say in hindsight: "Historians (perhaps even Mayr and Provine) have overemphasized the unity achieved by the synthesis." See Mayr, E. (1993), "What Was the Evolutionary Synthesis?," *Trends in Ecology and Evolution*, 8: 31–4 (quotation on p. 32). Of course, this is not to say that Mayr would have been prepared to recognize the level of disunity that is now being revealed in recent analyses, but it shows that with the passage of time the presumed achievements reached under the "Modern Synthesis" began to look differently even to its founders.

For the rather disproportionate and ideologically driven celebrations surrounding the 100th anniversary of the publication of Charles Darwin's *Origin of Species* (1859), consult Smocovitis, V. B. (1999), "The 1959 Darwin Centennial Celebration in America," *Osiris*, 14: 274–323.

We must issue a *mea culpa* with respect to jumping onto the Darwin bandwagon, and the pomp and fanfare of the 150th anniversary of the publication of the *Origin of Species* (1859) in 2009. This took the form of a special issue (Guest Editor, R. G. Delisle) in the journal *Studies in History and Philosophy of Biological and Biomedical Sciences*, Vol. 42, Issue 1, 2011, under the title "Defining Darwinism: One Hundred and Fifty Years of Debate." This special issue emerged from a symposium of the same title held at the University of Lethbridge (Canada) between November 12 and 14, 2009. RGD is now a repentant scholar.

It would be an endless task to try to list all the events and publications commemorating the 150th anniversary of the publication of Charles Darwin's *Origin of Species* (1859) in 2009, especially considering that it also corresponded to Darwin's 200th anniversary of birth in 1809. One telling example of Darwin's mythic status is the Wikipedia site marking "Darwin Day," in which one reads: "The day is used to highlight Darwin's contributions to science and to promote science in general. Darwin Day is celebrated around the world."

A brief anecdote (RGD): During a recent symposium in the United States, I was approached by a distinguished evolutionist who expressed annoyance at the critical stance I and others had adopted with respect to the development of evolutionary biology. The scholar expressed genuine concern that these investigations might lend aid and comfort to a common adversary, the anti-evolutionists. No doubt this is a legitimate concern, but it does not constitute good grounds for shrinking from self-criticism: again, critical self-analysis is precisely what separates science from nonscience and dogmatic thinking more generally.

Some evolutionists have found themselves unable to resist the temptation of using Charles Darwin as a sort of cudgel with which to beat evolution's opponents, thus giving him a disproportionate role in the history of ideas. While personifying the myriad complexities and debates within evolution may make for good narrative, such evolutionary biologists have, wittingly or unwittingly, contributed to the erection of a mythical Darwin. See Ruse, M. (1982), *Darwinism Defended: A Guide to the Evolution Controversies*, Reading: Benjamin/Cummings; Dennett, D. C. (1995), *Darwin's Dangerous Idea: Evolution and the Meaning of Life*, New York: Simon and Schuster; Mayr, E. (2001), *What Evolution Is*, New York: Basic Books, 8–11; Dawkins, R. (2009), *The Greatest Show on Earth: The Evidence for Evolution*, New York: Free Press, vii–viii, 3–18, 402–4. With one foot planted squarely in the static worldview, Darwin makes for a rather poor role model when attempting to confront Creationists.

Even for a zoologist like Ernst Mayr (1904–2005)—a preeminent figure of the "Modern Synthesis"—evolutionary biology can take on an element of atemporality. Mayr's focus on a biology centered around the principle of divergence (speciation), the principle of gradation (*Natura non facit saltum*), and the interpolation of past evolutionary events from what can be observed today places him in the direct intellectual lineage of Charles Darwin's a-historical biology. Consult Delisle R. G. (2009), "The Uncertain Foundation of Neo-Darwinism: Metaphysical and Epistemological Pluralism in the Evolutionary Synthesis," *Studies in History and Philosophy of Biological and Biomedical Sciences*, 40: 119–32.

For discussion of the rise of the field of "paleobiology," which allows genuine explanatory room for information gathered from the geological past, consult Ruse, M. (1989), "Is the Theory of Punctuated Equilibria a New Paradigm?," in A. Somit and S. Peterson (eds), *The Dynamics of Evolution: The Punctuated Equilibrium Debate in the Natural and Social Sciences*, 139–67, Ithaca: Cornell

University Press; Sepkoski, D. and M. Ruse, eds (2009), *The Paleobiological Revolution: Essays on the Growth of Modern Paleontology*, Chicago: University of Chicago Press; Turner, D. (2011), *Paleontology: A Philosophical Introduction*, Cambridge: Cambridge University Press; Sepkoski, D. (2012), *Rereading the Fossil Record: The Growth of Paleobiology as an Evolutionary Discipline*, Chicago: University of Chicago Press; Turner, D. (2017), "Paleobiology's Uneasy Relationship with the Darwinian Tradition: Stasis as Data," in R. G. Delisle (ed), *The Darwinian Tradition in Context: Research Programs in Evolutionary Biology*, 333–52, Switzerland: Springer Nature.

The case of Stephen Jay Gould (1941–2002) is an interesting one with respect to the quest for genuine historical thinking in evolutionary science. Indeed, while he himself became a major advocate for the recognition of a profound historicity in evolution, and a key figure in the foundation of the field of paleobiology, he nonetheless entirely overlooked the fact that Charles Darwin's thinking lacked such historicity. As seen in our Chapter 2, Gould thought that Darwin was the first truly modern evolutionist, vigorously promoting "evolutionary contingency," that is, a truly open-ended evolutionary process. We have seen in this book why this could not be the case: Darwin enclosed evolution in a straitjacket with the aid of a long list of explanatory devices. In short, Gould failed to see that the deficiencies he correctly identified as applying to the science of evolution in general were equally true of Darwin. Gould was, perhaps, hoping to ground his own thinking in the mythical Darwin to some degree, an almost irresistible temptation. In a sense, the authors of this book are pursuing Gould's quest by attempting to apply the same objectives to our understanding of the development of the field of evolutionary biology since its inception in the nineteenth century. For the two faces of S. J. Gould described here, compare his view of Darwin and of paleobiology: Gould, S. J. and N. Eldredge (1977), "Punctuated Equilibria: The Tempo and Mode of Evolution Reconsidered," *Paleobiology*, 3: 115–51; Gould, S. J. (1980), "The Promise of Paleobiology as a Nomothetic, Evolutionary Discipline," *Paleobiology*, 6: 96–118; Gould, S. J. (1980), "Is a New and General Theory of Evolution Emerging?," *Paleobiology*, 6: 119–30; Gould, S. J. (1982), "Darwinism and the Expansion of Evolutionary Theory," *Science*, 216: 380–7; Gould, S. J. (1986), "Evolution and the Triumph of Homology, or Why History Matters," *American Scientist*, 74: 60–9; Gould, S. J. (1989), *Wonderful Life: The Burgess Shale and the Nature of History*, New York: W. W. Norton, 277–91; Gould, S. J. (2002), *The Structure of Evolutionary Theory* Cambridge: Belknap Press.

Notes

1. Huxley, T. H. (1860) "Darwin on the Origin of Species," *Westminster Review*, 17: 542.
2. Smocovitis, V. B. (1999), "The 1959 Darwin Centennial Celebration in America," *Osiris*, 14: 321–3.

Index

anagenesis 19–22, 29, 32, 39
analogies. *See also* convergent evolution 26–8, 31, 32, 41
anti-Darwinian theories 7

balance of nature. *See also* evolutionary equilibrium 107, 113
Barzun, Jacques 48
Beer, Gillian 5, 116
  Darwin's Plots (2009) 5, 116
Bowler, Peter J. 122
Brongniart, Alexandre 101
  Description géologique des environs de Paris (1825) 101
Buckland, William 68, 101
  Geology and Mineralogy Considered with Reference to Natural Theology (1836) 100
Butler, Samuel 48

Chambers, Robert 100
  Vestiges of the Natural History of Creation (1844) 100
cladogenesis. *See also* divergent evolution 22–4, 29, 39, 42, 52
competitive exclusion principle 91–4, 96, 99, 110, 115–16, 125, 140
cone of increasing diversity 73, 124, 139
convergent evolution. *See also* analogies 24, 26–8, 32, 40, 42, 45, 78–9, 80, 90, 95, 96, 127, 137, 138, 139, 142
Cuvier, George 68, 101
  Description géologique des environs de Paris (1825) 101
cyclical evolution or motion. *See also* evolutionary equilibrium 99, 102–5, 109, 116, 125, 127, 130, 140

Darwinism 2, 122, 131, 137, 150
Darwin Myth or Legend 2, 3, 141, 145, 147–9, 154–5, 156
Dawkins, Richard 2
  The Greatest Show on Earth (2009) 2

decimation model 72
Depew, David 71
divergent evolution. *See also* cladogenesis *and* principle of divergence
  as applied in modern evolutionary theory 23–4, 26, 32, 39, 42, 45, 122–4, 127, 137, 139, 140, 142
  as applied by Charles Darwin 52, 54–6, 69–70, 78, 83, 89–96, 99, 110–12, 115, 119, 123–8, 130, 132, 137, 139, 140, 142

Einstein, Albert 5
Ellegard, Alvar 4
evolutionary contingency or opportunism 38, 46, 47, 49, 57, 58, 63, 75, 89–91, 114, 123, 127, 137–9, 141, 156
evolutionary equilibrium. *See also* cyclical evolution or motion 97, 99, 102, 104, 105, 107–10, 112–16, 119, 125, 127, 130, 140
evolutionary stagnation 29, 32, 45, 94, 96, 110, 112, 115, 119, 125, 127, 128–30, 139, 140
Extended Evolutionary Synthesis 48

Foucault, Michel 85, 87
  The Order of Things (1966) 85, 87

Galapagos Islands 3, 129–30
Gould, Stephen Jay 1, 2, 47, 71–2, 156
  The Structure of Evolutionary Theory (2002) 1
  Wonderful Life (1989) 47
Greene, John C. 99, 101, 116, 117
Grene, Marjorie 71
Griffiths, Devin 4, 73

Haeckel, Ernst 77
Holmes, John 5, 9
Hutton, James 101, 103–6, 117
  Theory of the Earth (1795) 103

Huxley, Thomas Henry 4, 48, 84, 87, 109, 127, 145–6

Jablonka, Eva 2
Jefferson, Thomas 106

Lamarck, Jean-Baptiste 100
  Zoological Philosophy (1809) 100
Lamb, Marion J. 2
Levine, George 4, 9, 47, 118
  Darwin and the Novelists (1988) 118
Limoges, Camille 85, 87
Linnaeus, Carl von 104–8, 110, 113, 117, 118
  Oeconomia naturae (1749) 104
  Oratio de Telluris habitabilis incremento (1744) 104
  Politia naturae (1760) 104
Lyell, Charles 4, 68–9, 71, 101, 106–9, 117–18
  Principles of Geology (1830–1833) 68, 106, 117

Mayr, Ernst 2, 46–7, 142–3, 154, 155
  The Growth of Biological Thought (1982) 46
Meckel, Johann 77
Modern Synthesis 48, 123, 143, 147, 150, 152–4, 155
monophyletism 15–19, 26, 28, 30, 31, 32, 33, 40–1

Natura non facit saltum 112, 124–5, 142, 155
natural theology 8, 118
neo-Darwinism 2, 48, 49, 143
Newton, Isaac 2, 51, 101–6, 146, 150

Owen, Richard 4, 84, 87, 109, 127

parallel evolution, 24, 32, 39–40, 42, 78–9, 80, 81, 90, 96, 127, 138
polyphyletism 15–20, 26, 28, 31, 32, 33, 41, 111
Post-Darwinian Theory 49–50, 51, 55, 57
preformationism 74–5, 140
principle of divergence. *See also* divergent evolution 69–73, 83, 87, 111, 112, 123–5, 131, 132, 139, 142, 155
principle of gradation. *See also* Natura non facit saltum 112, 124, 139, 155

random evolutionary walk 32, 96, 127
Ray, John 106
reticulate evolution 24–6, 32, 39, 42, 45, 90, 95, 137, 139, 142
Richards, Robert J. 132

Serres, Étienne 77
Smocovitis, Vassiliki Betty 147, 154
sociology of scientific knowledge 4, 5, 9, 10
standard view about the *Origin of Species* 47, 51, 54–6, 57, 58, 64, 116, 123, 132, 137, 150
steady-state world or view 101–2, 103, 106, 107, 109, 115, 118

uniformitarianism 68–9, 70, 102, 105, 107, 118

Von Baer, Karl Ernst 77

Whewell, William 68, 71, 118
Wilson, Edward O. 1
  Sociobiology: The New Synthesis (1975) 1
Winsor, Mary P. 123, 132

Young, Robert M. 101, 105

www.ingramcontent.com/pod-product-compliance
Lightning Source LLC
Chambersburg PA
CBHW061840300426
44115CB00013B/2455